REMAKING THE LANDSCAPE

REMAKING THE LANDSCAPE

The changing face of Britain

Edited by Jennifer Jenkins

P
PROFILE BOOKS

First published in 2002 by
Profile Books Ltd
58A Hatton Garden
London ECIN 8LX
www.profilebooks.co.uk

10 9 8 7 6 5 4 3 2 1

A CIP catalogue record for this book is available from the British Library.

ISBN 1 86197 375 6

Typeset in Bembo by MacGuru
info@macguru.org.uk

Printed and bound in Great Britain by
Clays, Bungay, Suffolk

Contents

Part Three: NATIONAL ATTITUDES

Plates

Tables and figures

The contributors

David Banister BA PhD MLIT is Professor of Transport Planning at the University of London and Director of Research at the Bartlett School of Planning at University College London. From 1994 to 1996, he was Visiting VSB Professor at the Tinbergen Institute in Amsterdam. In 2000-01 he has been a Visiting Fellow at the Warren Centre in the University of Sydney. He has an international reputation as one of the leading UK researchers in transport and planning analysis. His research has been extensively reported in sixteen authored and edited books, some twenty research monographs and over two hundred papers published in journals or as contributions to books. He has been recently involved with several research projects for the EU on scenario building and sustainable mobility, and was the director of the major Economic and Social Research Council programme on transport and the environment.

David Cannadine FBA became Director of the Institute of Historical Research, University of London, in 1998. He is the author, co-author and editor of numerous books including *Lords and Landlords: The Aristocracy of the Towns, 1774-1967* (Leicester University Press, 1980); *Patricians, Power and Politics in Nineteenth-Century Towns* (Leicester University Press, 1982); *The Pleasures of the Past* (Collins, 1989); *The Decline and Fall of the British Aristocracy* (Yale University Press, 1990), which won him the 1991 Lionel Trilling Prize; *G. M. Trevelyan: A Life in History* (Harper-Collins, 1992); *Aspects of Aristocracy: Grandeur and Decline in*

Modern Britain (Yale University Press, 1994); and recently *Class in Britain* (Yale University Press, 1998); *History in Our Time* (Yale University Press, 1998); and *Ornamentalism: How the British Saw Their Empire* (Penguin, 2001). He is general editor of the Longman 'Studies in Modern History' series, the Penguin 'History of Britain' and 'History of Europe' series, and a member of the editorial board of *Past and Present*. He is a regular broadcaster on UK television and radio.

John I. Clarke is an Emeritus Professor of Geography at the University of Durham where he was also Pro-Vice-Chancellor and Sub-Warden. His main academic interests have been in population geography and demography, about which he has written and edited many books, particularly in connection with three international committees that he chaired on population geography and population and environmental issues during the 1980s and 1990s. He is a deputy lieutenant of County Durham, and amongst other roles has been Vice-President of the Royal Geographical Society, chairman of the North Durham Health Authority and chairman of the Community Fund's North East Regional Committee.

Bryn Green OBE BSc PhD is Emeritus Professor of Countryside Management at the University of London. He has been head of the Environment sub-department at Wye College, University of London, the Nature Conservancy Council's regional officer for southeast England and a lecturer in plant ecology at the University of Manchester. His work is mainly concerned with conservation planning and landscape restoration and management, with particular interest in agri-environmental policy and practice, specifically in the integrated management of downland and heathland habitat for food production and amenity. He has written *Countryside Conservation: Landscape Ecology, Planning and Management* (George Allen and Unwin, 1981 and 1985, and E. & F. N. Spon, 1996) and has co-authored *The Diversion of Land: Conservation in a Period of Farming Contrac-*

tion (Routledge, 1991), *The Changing Role of the Common Agricultural Policy: The Future of Farming in Europe* (Belhaven Press, 1991) and *Threatened Landscapes: Conserving Cultural Environments* (Spon Press, 2001). He has been visiting professor to universities in the USA, Norway and Denmark and is a vice-president of the Kent Wildlife Trust and vice-chairman of the Kent Farming and Wildlife Advisory Group, which he founded as one of the first county groups. He was formerly a Countryside Commissioner, a member of the England Committee of the Nature Conservancy Council, chairman of the Landscape Conservation Working Group of IUCN and of the Kent White Cliffs Heritage Coast Countryside Management Project and deputy chairman of the Council of the Kent Trust for Nature Conservation. He has been a member of the editorial advisory boards of *Landscape and Urban Planning*, the *Journal of Environmental Planning* and the *International Journal of Sustainable Development and World Ecology*.

Ian Hodge is Reader in Rural Economy and Head of the Department of Land Economy in the University of Cambridge. He has published widely in the areas of environmental management, rural development and land use, including *Rural Employment* (with Martin Whitby; Methuen, 1981), *Countryside in Trust* (with Janet Dwyer; John Wiley, 1996) and *Environmental Economics* (Macmillan, 1995). He is a member of English Nature's socio-economic advisory group and was a member of the Department for the Environment, Food and Rural Affairs (DEFRA) Task Force for the Hills. He is currently leading an evaluation of agri-environment schemes for DEFRA.

James Hunter grew up in Argyll and lives near Inverness. A freelance writer and historian, he helped to found the Scottish Crofters Union and over many years campaigned for the land-reform programme that Scotland's parliament is now implementing. He chairs the board of Highlands and Islands Enterprise, the development agency for the north of Scotland,

and is a member of the board of governors of the prospective University of the Highlands and Islands. His books include: *The Making of the Crofting Community* (1976 and 2000); *A Dance Called America: The Scottish Highlands, the United States and Canada* (1994); *Last of the Free: A Millennial History of the Highlands and Islands* (1999); and *Culloden and the Last Clansman* (2001).

Dame Jennifer Jenkins is one of Britain's leading landscape campaigners. She is a former chairman of the National Trust and president of the Ancient Monuments Society. Her book, *From Acorn to Oak Tree*, on the history of the National Trust, was published (with Patrick James) in 1994. Earlier, she was for eleven years the chairman of the Consumers' Association. She has also been chairman of the Historic Buildings Council for England, the Royal Parks Review Group and the Architectural Heritage Fund. She is married to Roy Jenkins and lives in London and Oxfordshire.

Simon Jenkins writes a twice-weekly column for *The Times* and a weekly column for the London *Evening Standard*. He has written books on politics and on the history and architecture of London. He was educated at Mill Hill School and St John's College, Oxford. He began work in journalism on *Country Life* magazine and then worked for *The Times Educational Supplement* and the *Evening Standard*, edited the Insight page of the *Sunday Times*, and was editor of the *Evening Standard* from 1976 to 1978. He was political editor of *The Economist* from 1979 to 1986 and went on to found and edit the *Sunday Times* Books section, where he also wrote a weekly column. He was editor of *The Times* from 1990 to 1992. He has served as a board member of British Rail (1979–90) and London Transport (1984–6). He was deputy chairman of English Heritage from 1985 to 1990 and was voted Journalist of the Year in 1988 and Columnist of the Year in 1993. He was chairman of the Independent Commission on Local Democracy (1994–5) and was a member of the Millennium Commission (1994–2000). He is a member of the Buildings Books Trust and chaired the Booker Prize in 2000. His latest book, *England's Thousand Best Churches*, was published by Penguin in October 1999. He lives in London.

Richard Keen has spent much of his life in Wales, researching, writing about and lecturing on many aspects of its heritage. He is actively engaged in a wide range of preservation and conservation projects, and sits on a number of committees dealing with heritage in Wales. He is currently involved with a project mapping the cultural landscapes of Wales as well as writing a major guidebook on the country. He has published widely and written and presented many television and radio programmes. He has worked for the County Museum in Pembrokeshire and the National Museum of Wales, and for the National Trust as its Welsh Landscape and Culture adviser. He currently works as a freelance consultant.

Uwe Latacz-Lohmann completed his PhD in agricultural economics at the University of Göttingen, Germany, in 1992. He subsequently spent two years in the US at the University of Illinois and the US Department of Agriculture. He lectured in environmental economics at Wye College, University of London, between 1995 and 2000, and subsequently joined the University of Cambridge as Gurney Lecturer in the Department of Land Economy. He also holds an adjunct appointment at the University of Western Australia. His research interests are in agri-environmental policy, organic agriculture, environmental valuation and farm management. He has recently completed an economic evaluation of the Organic Farming Scheme for DEFRA.

Philip Lowe has been Duke of Northumberland Professor of Rural Economy and Director of the Centre for Rural Economy at the University of Newcastle upon Tyne since 1992. Between 1974 and 1989 he was successively lecturer in countryside planning and reader in environmental planning at the Bartlett

School, University College London. He is a specialist in the rural economy, and his research interests include the sociology of rural development, environmental policy analysis and land-use planning. He has authored and edited twenty books and over a hundred scientific papers. During 1997 and 1998 he was a member of the Minister of Agriculture's independent advisory group. He has recently been adviser on the rural economy to the House of Commons Select Committee on Environment, Transport and Regional Affairs; a member of English Nature's Socio-Economic Advisory Panel; an adviser to the European Foundation for the Improvement of Living and Working Conditions (an EU think-tank) on rural sustainable development in Europe; and an adviser to the European Court of Auditors. He is a board member of the Countryside Agency; a member of the Economists Panel, DEFRA; and chairman of the Market Towns Advisory Forum.

Oliver Rackham OBE is a Fellow of Corpus Christi College, Cambridge, and a botanist and historian of vegetation and landscape, especially woodland. He has observed the woods around Cambridge since 1958. Other areas of interest include Wales, southern Europe (especially Greece), North America, Japan and Australia. His objective is to combine studies of the behaviour of plants and animals with evidence derived from written records, archaeology, pictures and photographic archives, and the materials of ancient buildings. His OBE was awarded for 'services to nature conservation'. His books include: *Trees and Woodland in the British Landscape* (Dent, 1976; second edition, 1990); *Ancient Woodland: Its History, Vegetation and Uses in England* (Edward Arnold, 1980; second edition forthcoming); *The History of the Countryside* (Dent, 1986); *The Last Forest: The Story of Hatfield Forest* (Dent, 1989); *The Illustrated History of the Countryside* (Weidenfeld & Nicolson, 1994); with A. T. Grove, *The Nature of Southern Europe: An Ecological History* (Yale University Press, 2001).

Marion Shoard started her working life as a scientist, took a town-planning qualification in order to work in conservation, and joined the Council for the Protection of Rural England as its first planning specialist in the mid-1970s. In 1980, her book *The Theft of the Countryside* first drew attention to the damage being done to the English countryside by modern farming, and set out proposals for conserving and rehabilitating the landscape. Her second book, *This Land Is Our Land*, on the nature of rural land ownership, was published in 1987 and put forward radical proposals including a rural land-tax-and-grant scheme; this book was updated, expanded and re-issued as a Gaia Classic in 1997 (Gaia Books). In her latest book, *A Right to Roam*, published in 1999 by Oxford University Press, she explored the historical and philosophical basis of rights of access to the countryside and put forward detailed proposals for a universal right to walk; this book received the Outdoor Writers Guild's Best Outdoor Book of the Year Award in 2000. She is currently writing a guide for relatives to the care of elderly people, after her 87-year-old mother developed Alzheimer's disease and became blind.

Sir Crispin Tickell GCMG KCVO is Chancellor of the University of Kent at Canterbury; chairman of the Climate Institute of Washington DC; and president of the Earth Centre in South Yorkshire. Most of his career was spent in the diplomatic service. He was chef de cabinet to the president of the European Commission (1977–80), ambassador to Mexico (1981–3), permanent secretary of the Overseas Development Administration (1984–7) and British permanent representative to the United Nations (1987–90). He then became Warden of Green College, Oxford (1990–97), and remains director of the Green College Centre for Environmental Policy and Understanding. He was president of the Royal Geographical Society (1990–93), chairman of the International Institute for Environment and Development (1990–94), chairman of the government's advisory committee on the Darwin Initiative (1992–9), president of the

Society for Clean Air (1997–9) and convenor of the government
Panel on Sustainable Development (1994–2000). He is author of
Climate Change and World Affairs (1977 and 1986), and *Mary
Anning of Lyme Regis* (1996). He has contributed to many books
on environmental issues, including human population increase
and biodiversity. He is a member of two government task forces,
one on urban regeneration and the other on near-earth objects.
His interests range from business and charities to climate, mountains, pre-Colombian art and the early history of the earth.

REMAKING THE LANDSCAPE

Introduction

Jennifer Jenkins

The one thing that is certain about British landscapes is that they will change, and change dramatically, during the next twenty years. This book has its origin in fears that such changes will inevitably be for the worse, as have been many during the last fifty years. But in fact most of the authors believe that there is now a chance of making changes for the better in both town and countryside.

Why is there this hint of optimism despite the effects of climate change, the recession in agriculture and the explosion in travel by car throughout Britain, accompanied in the south and east by the pressures of an increasing population and in parts of the north and west by economic decline? Paradoxically it stems from the very failures of present policies and the new thinking that they have generated. A vision of how our towns and cities could become places where people want to live rather than to leave has been set out in the Urban Task Force's 1999 report 'Towards an Urban Renaissance', with its goal of creating the 'Sustainable City'. The concept of sustainable development offers the possibility of making travel less necessary and less dependent on the car. And foot-and-mouth disease, the latest disaster to hit farming on top of dissatisfaction with the system of subsidies encouraging intensive output, will make it necessary to rebuild the rural economy on a more diverse basis, which could yield a countryside far richer in wildlife.

A landscape is defined by the *Oxford English Dictionary* as 'A view or prospect of natural inland scenery, such as can be taken in at a glance from one point of view.' In this book the word is used also to include urban views, whether beautiful or not. It covers such places as the view of Whitehall from St James's Park, 'a composition of landscape architecture unsurpassed in the world',[1] and the prospect of Edinburgh Castle from Princes Street as well as the edgelands at the interface between town and country where the most obvious sights are rubbish tips, crushed cars and warehouses, cheek-by-jowl with derelict industrial sites. Rural landscapes are no less diverse, ranging from the Highlands and Islands of Scotland and the peaks of Snowdonia through the 'sodden and unkind'[2] Midlands to the potato fields of the Fens, the cornlands of East Anglia and the swelling downs looking across the English Channel.

It is not possible to describe a landscape as precisely as a building, with its exact dimensions and architectural features. The composition of a view becomes larger or smaller as one approaches or goes further away – the view of St Paul's Cathedral is much more extensive when seen from Greenwich than when seen from the Tate Modern. This makes it even more difficult to protect an urban landscape than a building. Any attempt can arouse opposition from architects and developers. The only measure introduced to safeguard the historic skyline of London is that protecting ten views of the Palace of Westminster and St Paul's from the surrounding hills and other vantage points, and even this is contested by the promoters of high buildings which would obscure the sightlines. Rural landscapes, apart from carefully laid-out estates, present still more difficulties of definition, and of protection, unless within a national park.

This book is concerned with ordinary landscapes throughout Britain rather than with the most beautiful and the most historic. The latter were the first concern of campaigners and many of these are now reasonably well protected. National Parks and Sites of Special Scientific Interest were designated in England and Wales after 1949, since when they have increased in number and scope. It was another twenty years before effective measures were brought in to stem the demolition of historic buildings, soon followed by protection for conservation areas. Protection through planning control has been added to, and in some respects led, by the much stronger control of protection through ownership. Over a century ago, the National Trust set out to preserve the coastline and now protects over 600 miles (965 kilometres) in England and Wales. The Trust was also a pioneer in saving historic country houses from demolition or decay, introducing a rescue scheme as early as 1937 and another for gardens ten years later. The National Trust for Scotland has played a similar, but not identical role. In the absence of national parks there, it has acquired mountainous land and has specialised in promoting the restoration of small houses.

The interdependence of town and country has been recognised since planning first became a political issue. Only if we can make our towns pleasant places in which to live, enhancing the inner residential areas and the suburbs as well as the central public squares, will it be possible to slow down, let alone to halt, the gradual suburbanisation of the countryside. At the end of the Second World War, effective planning legislation was thought to hold the key to the protection of the rural landscape. The system has had some success in confining most new building to existing towns and their outskirts, an achievement that contrasts with what has happened in many sought-after rural areas of France. But successive planning acts have not been able to prevent the relentless urban advance by nearly 13 square miles (33 square kilometres) a year in England between 1994 and 1998. The ambitious plans for post-war reconstruction produced some architectural icons, but far more abundant were the high-rise flats so unpopular with their tenants that many have been demolished. Few new urban settlements have been created worthy to pass on to future generations, although some of the 'new towns' are exceptions.

Dreams for outdated industrial areas, such as that for the south bank of the Thames from County Hall to Greenwich, have mostly been stillborn. Maynard Keynes, in the 1937 collection of essays *Britain and the Beast*, thought that this stretch of river could become one of the sights of the world, with the existing population housed in half the area and the rest being devoted to 'parks, squares and playgrounds, with lakes, pleasure gardens, and boulevards, and every delight which skill and fancy can devise'. The vision of a new cultural centre stretching east from the Royal Festival Hall has, however, gradually been realised, enabling the arts to jump the previously rigid frontier to 'capital city' London. But the blocks of flats built during the last two decades are squeezed up against the river from which they are separated by a narrow path rather than by tree-lined boulevards. These desirable residences for the affluent form an almost continuous wall without parks or gardens to reveal river views to the less affluent people living behind. Individual buildings, however striking in design, are not enough to create an attractive city for all its inhabitants unless set in green surroundings and in harmony with their neighbours. What a contrast this is to the Old Royal Naval College, Greenwich, London's masterpiece of architectural landscape, where the succession of great buildings leads the eye up to the park beyond.

Newcastle and Gateshead have been more successful in opening up the banks of the Tyne. Formerly cut off by wharves, the river is now lined with gardens, trees and well-mannered new buildings, alternating with older buildings of different periods in use for a variety of activities, the two banks linked by an exciting new bridge. Perhaps a less buoyant economy and lower land values make it easier to create a more generous layout and to allow more public spaces.

Anyone involved in managing urban or rural environments will have to make provision for the likely effects of climate change, as outlined by Crispin Tickell. Some of these effects are already being seen. In East Anglia, for example, the coastline is receding in some places and being added to in others, while inland some areas are reverting to conditions before the Fens were drained. Not only farmland is being lost but also natural habitats on salt marshes and mudflats, though these may be partially replaced by accretions further along the coast. It is essential that rising sea levels should be anticipated and managed rather than resisted, as King Canute would have tried to do. Attempts to hold back the overwhelming strength of the sea along the whole length of the coast would be excessively expensive and in the long run futile. Where physical barriers are needed to protect existing towns they must be sensitively designed: as the Thames Barrier shows, they can make new and dramatic landmarks.

The predictions of higher rainfall for some parts of Britain are likely to bring more frequent and destructive flooding, such as was seen during the winter of 2000/01, unless countervailing steps are taken. Environmentally the most desirable means of dealing with higher rainfall would be to allow upstream stretches of river to return to a natural meandering course and gently to flood uninhabited water meadows. Farmers could be paid for the loss of production and for creating natural habitats. If, as Tickell suggests, the north and west become wetter, trees will grow more quickly, and if the south becomes drier, crops may have to be varied to suit conditions more like those of the Mediterranean.

More subject to human control are the pressures arising from the growth, migration and composition of the population described by John Clarke. Already one of the most densely inhabited countries in the world, Britain is expected to increase in population at the rate of about 2 million a decade during the next twenty years, as it has done during the last fifty. The population increase will be concentrated in the south-east, while most of Scotland, Wales, and the north-east and north-west of England will remain stable, a trend that has been fairly persistent since the end of the First World War and is likely to continue. Some parts, however, have experienced an actual decline. The uninhabited or very sparsely inhabited areas of Britain occupy half the land but contain only 1–2 per cent of the population.

Another major demographic trend examined by Clarke is the ageing of the population; 16 per cent of British citizens are now 65 or over and the proportion will continue to rise. This, together with the decline in the size and stability of the nuclear family and the increase in total population, has already caused the number of households to increase by 5.7 million since 1961; a further 3.8 million extra are expected in England by 2021. Effectively the increase comes from the over-45 age group, of whom almost a third are living alone.

Since the 1960s there has been a continuing exodus from the cities to the suburbs, the smaller towns and the rural areas, especially in southern England. At the same time, shopping, leisure and employment have been decentralised. There was, however, some slowdown in the number of people leaving the cities during the 1990s and this is expected to continue for the next decade, though with the usual regional differences. London's population is expected to rise by 7 per cent because of immigration, compared with an average decline of 1 per cent in metropolitan areas generally. Yet even where the conurbation as a whole has declined, the city centres, such as Liverpool, have gained in population. As the author shows, the changing geography has been accompanied by a massive increase in the stock of dwellings, from 13.8 million in 1951 to 24.6 million in 2000 and by an increase in owner-occupiers from 29.6 per cent to 67.8 per cent, adding to the pressure for building on greenfield sites (virgin land on which there has been no previous development).

The pattern of development is determined by how much people travel as well as by the size and location of the population. David Banister traces these changing travel patterns, in a country where the average distance travelled each year is estimated to have increased by 28 per cent since 1985 while the number of individual trips has remained the same. Most of this increase has been due to the growth in the use of the motorcar, which by the end of the century accounted for the majority of journeys and four-fifths of the distance travelled. Walking, cycling and public transport have all declined.

Millennium year did not augur well for righting the balance.

Outfaced by a militant roads lobby, the government cancelled the automatic escalator in petrol prices and increased the investment planned for roads; the railways fell into chaos; and the London Underground became almost insupportable while plans for its upgrading remained in dispute. The only significant step forward was the national cycle network, initiated by an independent charity and part-funded by the Lottery. If this were supplemented by local networks of cycle tracks, separated from vehicles and linking homes, schools, jobs and stations, more people would be tempted to leave their cars in the garage. Still more would do so if public transport were raised to the best continental standards and fares were reduced to their levels.

The Urban Task Force report describes the concept of an 'urban renaissance' as being about 'creating the quality of life and vitality that makes urban living desirable'. This will be achieved only if cities are specifically planned on the lines put forward by Banister. Housing, shopping and employment should be located close together, near public transport and within easy reach of a full range of services. This will involve reversing the policy of concentrating hospital and other services on a smaller number of larger sites. The disadvantages of longer travel must be set against the loss of somewhat problematical economies of scale. The popularity of such a move was illustrated at the 2001 general election by the overwhelming vote in favour of retaining the facilities of the Kidderminster local hospital. Bringing services nearer to where people live would contribute to sustainable development, an aspiration still more honoured in rhetoric than in action.

What prospect is there of attaining this urban renaissance, desirable in itself and necessary to lessen the exodus from the cities to the countryside? The changing structure of the population, particularly the increasing number of one- and two-person families, will improve the prospects for higher-density living. For these people, there is a lot to be said for the city's leisure resources and the opportunity to save time and money on commuting. The suburbs, always more popular with family homebuyers, also need housing specifically designed to suit the needs of people living alone. They could then attract a wider range of activities and public services and become entities in their own right, linked together by better public transport.

The importance of design to urban regeneration can be seen in Barcelona. The former mayor of Barcelona, Pasqual Maragall, who presided over the city's revival, set out to repair the damage to areas that had fallen into decay, recycling old buildings and introducing a strict historic preservation code, while at the same time clearing the obsolete docks and opening the sea to the public via new promenades and parks. Run-down areas away from the centre were not neglected, their public amenity improved by well-designed spaces and attention to the detail of the streetscape. The Olympic village was built on the site of a derelict industrial area on the coast and it is here that two 42-storey skyscrapers were built, not interrupting the historic skyline, but forming a distinctive new prospect.

The keys to success in Barcelona were much the same as those identified by Simon Jenkins as necessary if Britain is to create towns that could rival Perugia, Amsterdam or Geneva. He sees the biggest challenge as being that of design, with revived civic leadership an essential precondition and planning the only way to proceed. Jenkins, architectural historian of range and verve, is pessimistic about the outcome; his gloom is fortified by the crude monstrosities of the new centres of Aylesbury and Basingstoke and the earlier tall buildings that obstruct the vistas of Newcastle and Bristol. Richard Rogers, architect of the Pompidou Centre in Paris and chairman of the Urban Task Force, is on record as having more faith in the capacity of professionals to improve the quality of urban design.

The government accepts Jenkins's objectives to the extent of setting up the Commission of Architecture and the Built Environment to promote higher standards of design. Public buildings are expected to set an example. However, the commercial pressures imposed by the Private Finance Initiative too often mean that profit takes precedence over quality of design and materials.

Some cities have benefited from Millennium projects of outstanding quality that are not subject to such pressures, but even the most spectacular new buildings do not help to create good neighbourhoods unless they are knitted into the fabric and are set in well laid-out public spaces. The Lowry Centre in Salford demonstrates how a landmark building can be a focus for a new local environment.

A desire to achieve the 'revived civic dignity and leadership' called for by Jenkins has prompted the government to introduce the opportunity for mayors to be elected. But procedural changes will not have much effect unless the Treasury is willing to loosen the financial reins and give local authorities freedom to raise the money needed to enhance the environment and improve other services. The refusal to allow the Mayor of London to decide what was the most efficient way to run London Underground before he took responsibility for its management is just one example of the reluctance to delegate responsibility. Joseph Chamberlain would never have been able to make Birmingham the best-governed city in the country during the 1870s had his powers been so restricted: he would not have had the opportunity to acquire the local gas works, to buy the Elan Valley in Wales as a source of pure and plentiful water, or to develop the town as a thriving business centre yielding a valuable income for the council.

It will not be easy to revitalise those towns and cities in the north and west that are suffering from the decline of manufacturing and heavy industry. Despite all the efforts of successive governments since the mid-1930s to redistribute employment, the north–south divide in economic performance has persisted. A 1999 report from the Town and Country Planning Association does not expect present regional policies to have more than a marginal effect. Nonetheless, the two capital cities of Edinburgh and Cardiff are prospering, as are the cathedral cities of York and Chester. Jenkins highlights the revival of Victorian industrial quarters in cities such as Leeds, Birmingham and Manchester, which have become popular for living and leisure.

In these cities, historic areas that thirty years ago were becoming derelict have been restored and their beauty has again become apparent. Yet economic decline does not necessarily bring a more compact city. Newcastle, for example, has plenty of empty houses and derelict land, but still wishes to build on greenfield sites.

The results of weak planning control are seen at their most extreme in what Marion Shoard terms the 'edgelands', the interface separating town from countryside. 'Often vast in area, though hardly noticed, [this interface] is characterised by rubbish tips and warehouses, superstores and derelict industrial plant, office parks and gypsy encampments, golf courses, allotments and fragmented, frequently scruffy, farmland. All these heterogeneous elements are arranged in an unruly and often apparently chaotic fashion against a background of unkempt wasteland frequently swathed in riotous growths of colourful plants.'

Shoard outlines the history of the edgelands, which took off during the Thatcher years, when one new superstore could trigger an explosion in development. In the Isle of Thanet, for example, the initial planning application for land at a junction between main roads leading to Margate, Broadstairs and Ramsgate was turned down by the local authority but was allowed on appeal. After that the local authority felt obliged to grant further applications and was unable to control the ensuing sprawl of self-contained retail parks and advance-built factories, orientated to car-based travel and sucking life from the now unfashionable seaside resorts.

How then should the edgelands be dealt with? They are useful, accommodating disposal plants, sewage works and other essential if unattractive services as well as offering unexpected places for wildlife to flourish, children to play and teenagers to test their motorcycles. But their isolated blocks of housing and much larger areas of employment lack basic services, let alone public parks. They cannot continue to proliferate haphazardly, in response to planning applications, without detailed land-use plans. For Shoard, the edgelands present an opportunity for planners to exercise their skills, converting fragmented housing into urban villages, linking the business parks to the public transport system and creating networks of green spaces, yet without losing the untamed character of the interface.

Some edgelands are situated within green belts, the oldest and most popular planning designation. Still serving the original purpose of preventing the coalescence of neighbouring towns, green belts involve longer travel distance as people seek to buy houses beyond their borders and then commute back to work. For this reason, they are now being challenged by those who consider that the good of sustainable development would be better served if new building were permitted along transport routes penetrating their boundaries. A far better alternative would be to create more compact communities and thus reduce the need to travel. Ribbon development along transport routes is the opposite of what is required.

Agriculture occupies 80 per cent of Britain today and thus largely determines the state of the countryside. How farmers manage their land is crucial to the landscape. An eminent ecologist, A. G. Tansley, writing in 1946 concluded that the reclamation of uncultivated land during the Second World War 'has not on the whole diminished the beauty of the countryside – rather the contrary is true . . . because it has replaced monotonous and often practically deserted and derelict grass-fields by the varied and stimulating activity associated with plough-land'.[3] He went on to say that farming should continue except in areas of special scenery and native vegetation, which should be protected. He did not foresee the subsequent intensification of agriculture, which would eliminate hedgerows and ancient woodlands, moors, heaths and water meadows and, with the increasing use of chemicals, would devastate wildlife and diminish beauty. Even today many people would be shocked by Bryn Green's judgement that farming constitutes the major force destroying countryside amenities and has eliminated much of its biodiversity and archaeology.

Britain's agriculture ranges from livestock farming in the uplands of the north and west to arable cultivation in the east and south-east. The two have increasingly diverged as marginal grazing land has been managed less intensively, afforested or abandoned and the better land has been devoted to more intensive crop production. Ian Hodge and Uwe Latacz-Lohmann show that it is in the lowlands that farming has had the most impact, partly because the proportion of land in arable production has such a visible effect on the landscape. This proportion has varied greatly, falling to a low of 30 per cent in the late 1930s, then rising to peaks of almost 50 per cent during the Second World War and again in the early 1980s. Since then there has been another decline. The first peak was caused by the need to maximise production during the war and the second by the effect of the Common Agricultural Policy (CAP) after Britain entered the European Union.

Since 1984 measures have been introduced to restrain production and divert CAP subsidies to agri-environmental schemes. In the intensive arable areas the principal instrument has been the compulsory set-aside scheme. This has brought some benefits for wildlife, mainly through the creation of buffer zones, but the unkempt, uncultivated fields do not add to the beauty of the countryside.

Hodge and Latacz-Lohmann predict that future reforms to the CAP will almost certainly include cuts in production subsidies and in the long run perhaps their complete elimination. But even if returns to farmers are less than during the past 50 years, arable production is likely to continue to be commercially viable on good-class lowland acres, provided that a reduction in subsidies is matched in other countries. These areas will be dominated by bulk producers, who are under pressure to reduce costs. They will adopt new practices of precision farming, making it possible to reduce the input of chemicals, and will abandon small awkward fields and other less profitable land. A few farmers will turn to organic farming or production for niche markets and this will add variety to the landscape in the relatively small lowland areas where it is likely to be adopted. Wetland areas are already being planted with alder, poplar and willow, for which there is an industrial market. If land prices fall, as is likely, environmental organisations may be able to purchase more land for conservation, but any benefit to the landscape would be offset by the extension of fenced plots owned by individual houseowners.

If subsidies continue to be available in the arable areas they are likely to be conditional on the provision by farmers of wider benefits. Hodge and Latacz-Lohmann suggest two schemes that, if applied to perhaps 10 per cent of the lowlands, would make an appreciable difference in the most intensively farmed parts of the country. The first would be a modified set-aside scheme under which farmers would bid for grants to take land out of production and manage it for such purposes as nature reserves, buffer zones for watercourses or public access. Bids could be concentrated where they offer relief from the most intensive agriculture and are within reach of large centres of population. The second scheme would benefit declining species associated with arable farmland by offering payments for such measures as retaining stubble during the winter, extending field margins or sowing mixtures of wild seed. Schemes of this sort, open to all farmers who wish to participate, would do something to foster wildlife habitats, enliven the vistas of large uniform fields and compensate farmers for the end of production subsidies.

In the arable areas production is likely to be on an increasing scale in order to maximise profits and to be carried out by large highly technical organisations with little or no local associations. Some owners already contract out a major part of their operations to large companies, which may also buy or rent land directly. In the livestock areas the contract system is also becoming increasingly common and here it often involves local farmers. The association between ownership and management is weakening as older farmers retire from an active role and non-farmers buy properties for recreational and residential use.

In the countryside as a whole there has been an increase in

farm owner–occupation from barely 10 per cent in 1914 to 48 per cent in 2000. Among small farmers the intensity of operation varies widely, but the authors point to the absence of any real evidence to support the commonly held view that 'family' farms are in general better for the environment than are large ones. A few small farms remain as relics of a past age, but more often it is farmers who enjoy outside incomes and managers of larger businesses who give space to wildlife and have the labour to maintain hedges, walls and other landscape features. In the case of tenant farmers much depends on the policy of the landlord. The National Trust, for example, which owns large scenic areas, frames its leases to require good conservation practices in return for a lower rent.

In the uplands of the north and west of England agriculture is no longer the mainstay of the rural economy. Tourism employs more people and contributes more income, as it does in many mountainous and coastal areas of the Continent. Yet the countryside is still shaped by farming and, as the outbreak of foot-and-mouth disease in 2001 showed, is vulnerable to an agricultural crisis. Philip Lowe traces the development of the crisis from its inception. He argues that agriculture has symbolic and aesthetic benefits that far outweigh its marketable benefits in the production of food – visitors are attracted to the Lake District, the North Pennines and similar areas because of their image as pastoral landscapes maintained by extensive grazing. The failure of the government to recognise this, treating the outbreak purely as a question of animal health, i.e. an *agricultural* problem, discouraged people from visiting the countryside. This in turn precipitated a crisis in the wider rural economy as hotels, restaurants and leisure facilities were bereft of customers. In contrast, the outbreak of swine fever in East Anglia the previous year, although serious for that area's pig sector, did not have similar ramifications for the wider rural economy, as East Anglia is a region less dependent on the tourist trade.

Foot-and-mouth disease struck when the livestock industry was already in deep depression and incomes were at their lowest level for a generation. A large part of upland Britain was denuded of its grazing herds and it is not clear how far they will be replaced. Lowe predicts that some farmers will retire early and dispose of their land, and others will not fully restock. There are likely to be more amalgamations and 'ranching', accompanied by less intensive shepherding as well as more part-time and hobby farming. Large areas may be abandoned and revert to woodland, as have vast tracts in the north Mediterranean and in eastern North America. Scrub and tree encroachment is in fact already far advanced on most surviving heath and downland, and is appearing on the northern hills and moors.

However distressing it may be to see the abandonment of land that has been farmed for generations, it would not, in Bryn Green's words, 'be desirable for the entire countryside to be made into a museum of ersatz landscapes, with a hundred years' worth of technological progress in agriculture abandoned and traditional techniques frozen in an unchanging time warp'. It would, however, be unacceptable for the Lake District and other historic landscapes that are celebrated in art and literature to be taken over in large part by woodland. As well as a great cultural loss, this would be a severe blow to the tourist trade.

Philip Lowe calls for a strategy for the uplands that balances the conservation of the landscape and its traditional farm buildings with the economic forces for change. This should be the starting point for a recovery programme in the livestock areas based on a 'territorial policy for rural development' as a whole rather than on a sectoral policy for agriculture alone. Such a programme would have benefits for farm households, the majority of whom have some non-farming income – outside work produces more income than bed and breakfast and other tourist enterprises based on the farm. In fact, more than two-thirds of people in rural areas have employment in towns that are reasonably accessible to the four-fifths or more of rural households that own a car. Lowe also urges that subsidies, which since the Second World War have made possible the survival of livestock farming in the hills, should be detached from output and directed to environmental land management. In other words the production side of farming should operate as a business dependent on market prices, and the work in maintaining landscapes, protecting wildlife and providing access should be regarded as a public service to be paid for under the CAP and other public funding.

As Bryn Green shows, Britain has made less use of the CAP's agri-environmental schemes than have most other European Union countries. The government, unwilling to pay the national contribution required, has allocated only £1.6 billion over six years for these schemes, compared with the £3 billion each year for production subsidies. If farmers were no longer able to depend on flat-rate payments from the public purse for every animal they produce, they would be more likely to respond to the growing consumer demand for high-quality food from a known source and for organic produce. Since the demand for organic food is being largely met by imports, it would make sense for more financial and technical help to be given to the lengthy process of conversion. Advice is also needed on how to process and market locally distinctive products on a village or regional basis in the same way as varieties of cheese are sold in France and as local specialities are being developed in Germany.

For the rural economy to prosper, market towns need to be nurtured as centres of employment, training and services for the surrounding rural areas. Philip Lowe describes how some of these towns are thriving and gaining population as more people seek the lifestyle they offer, while others are suffering from the loss of trade to out-of-town developments and the concentration of services in larger centres. Economic decline is often accompanied by a down-at-heel appearance and a poor welcome for visitors. Well-designed and maintained public squares and gardens, conveniently located car parks and good sports facilities make an immediately favourable impression, as can be seen in many French provincial towns.

The retreat from farming will offer opportunities for new

landscapes to be enjoyed by a population that has more time for leisure. Hundreds of golf courses were laid out on retired farmland during the 1990s and although they undoubtedly suburbanise the landscape, they are better kept than the litter-strewn fields in the urban fringe and generally richer in wildlife than cultivated land. More unusual is the park constructed from spoil from the Channel Tunnel below the cliffs of Dover, already colonised by thousands of rare orchids. Most extensive of the new landscapes will be the woodlands. Oliver Rackham describes how in the 1950s and 1960s many ancient woods were replaced by either arable land or conifer plantations, before conservationists made their voices heard in the 1980s and landowners began once again to understand and take pride in their woods. He predicts that the difficulties being experienced by the agriculture industry will lead to much former farmland reverting to woodland, as has already happened in Essex and other parts of the country.

Rackham expects the woodlands in time to become 'a mosaic of natural recent woods and plantations, embedded in which will be the existing woods and trees'. The new woods, which deer, sheep and the grey squirrel will play their part in shaping, will not be the same as the existing ones, nor will they be a replacement for ancient woodland that has been destroyed. 'Their ground vegetation will not be bluebells and primroses, but the nettles and brambles of fertiliser-sodden soil,' but they could, if skilfully managed, become attractive places for walking, riding and peaceful relaxation.

In the Highlands and Islands of Scotland, as in Wales, landscapes are seen from a very different point of view than are those in the south of England. In these countries landscapes reflect the long history of peoples who have endured conquest, loss of culture and depopulation and who are now trying to reassert their national identity.

James Hunter evokes the beauty of the Highlands and Islands, the least densely populated area of the European Union outside arctic and subarctic Scandinavia, tracing their history and suggesting strategies for their future prosperity. Many glens and Hebridean islands contain fewer people than they did five thousand years ago. Traces of ruined houses show that communities were settled here for generations before they were forcibly removed about two hundred years ago in what are known as the Clearances. It is not surprising that many highlanders see their country not as a wilderness to be preserved but as a place where people could live once again and in so doing would complete rather than damage the often spectacular landscape.

Until recently the prospect of repopulation of the Highlands and Islands seemed little more than a dream. But in the last twenty years, when the population of Scotland as a whole has been static, the population of this region has risen by nearly 20 per cent and the number in employment by more than 40 per cent. Although much of the increase has been around Inverness, more peripheral areas such as Skye have also gained. The change is partly the result of the increasing self-confidence of Highland communities, shown by their growing pride in Gaelic culture and their recognition of the landscape as an economic asset as well as a place of extraordinary beauty. It is also due to technological advances that have enabled a wide variety of business activities to operate away from towns and to complement the inadequate income to be derived from hill farming.

For people making a living in the Highlands and Islands the traditional pattern of crofting communities, where each household occupies a few acres of land, offers a better prospect than does an estate on the edge of a village. For this reason there is a strong demand for smallholdings in places that have hardly been inhabited since the Clearances. Hunter is convinced from his experience as chairman of Highlands and Islands Enterprise that some of the land presently used for hill farming, which for decades has not been viable without subsidy, should be turned over to new crofting communities. This has already happened on the island of Eigg and on other estates in the west, but is inhibited by the concentration of land ownership in the hands of a tiny number of proprietors.

There is no reason why small new settlements should detract from the landscape, provided that they are visually as unobtrusive as their eighteenth-century predecessors, are fitted into the contours of the land and are built of local materials. If CAP grants were diverted from production to land management they could help with the extra cost of locally quarried stone as well as aiding the maintenance of landscape features and natural habitats.

In the Highlands and Islands conservation must go hand in hand with change. James Hunter sees 'a future in which people become once more an integral component of Highlands and Islands landscapes'.

The Welsh historian O. M. Edwards famously proclaimed, 'Our land is a living thing, not a grave of forgetfulness under our feet. Every hill has its history, every locality its romance, every part of the landscape wears its particular glory.' Richard Keen describes how this feeling for the landscape is expressed in the legends, literature and place names of Wales, which evoke the scene and sometimes even the prevailing weather of the land, as does Twyn-y-gwynt – hillock of the wind. The countryside illustrates the history of Wales from the Neolithic tombs and Roman settlements to the great castles of Edward I, the medieval abbeys and churches, and the few surviving country houses with their lush romantic parks and gardens. On the hillsides are scattered the small whitewashed houses of the tenant farmers, in some places now overshadowed by prominent and inappropriate modern villas.

Agriculture and forestry dominate the Welsh landscape as they have done for centuries. The landscape is fundamental to the prosperity of Wales – it has been estimated that in 1999 £6 billion of Welsh GDP was directly dependent on the environment and that its management accounted for one in six jobs. Wales is not a rich arable country, its mountains, thin soil and rainfall making it more suitable for livestock than for corn. But hill farming is suffering from the same economic difficulties and consequences of foot-and-mouth-disease as the north-west of

England and is likely to see similar changes. Keen predicts that some land will probably cease to be farmed and give way to woodlands, either planted or regenerated naturally. And if CAP subsidies on production are replaced by grants for stewardship of the countryside, the different emphasis could benefit the rural economy as a whole, including tourism.

Visitors have long been attracted to the coast of Wales: some would say over-attracted when they look at some monotonous developments. The serried ranks of caravans lining the coast of north Wales have provided accommodation for Liverpudlians in search of sea air, just as the caravan park on the outskirts of Porthcawl, once the largest of its sort in Europe, provided an escape for those from the coal valleys. More attractive is Tenby, its early nineteenth-century terraces set on a promontory with views along the coast in either direction.

The mining and iron and steel industries have also left their mark on the landscape, though almost all has now been swept away. Among those that survive, Keen cites the towering tips of the slate quarries in Blaenau Ffestiniog and, most interesting from a historical perspective, the succession of mining and iron works around Blaenafon, now designated a World Heritage Site. It is difficult to envisage that at its peak in the early twentieth century, 250,000 men were employed in coal mining and the valleys of south Wales were filled with slag heaps and the distinctive pithead winding gear that carried the men down to the seams hundreds of feet underground. Only fifty years ago it would have been inconceivable that these valleys should have been 'greened' by massive land reclamation schemes, that one of the handful of mines to survive should be a museum and that Blaenafon, with its grand Workmen's Institute and many chapels, should be a solitary reminder of the close-knit mining communities with their vigorous cultural and political life.

The closure of the mines has been accompanied by the closure of most of the docks that were built to export vast quantities of coal. The coastal strip from Newport to the west has been redeveloped for modern industries and a striking new landscape has been created within the Cardiff Bay Barrage. Contemporary hotels, offices and flats are to be joined by the Millennium Hall and a building for the Welsh Assembly, though whether the designs will be worthy of the setting remains to be seen.

These essays illuminate the opportunity, in David Cannadine's words, to reconcile 'development with conservation, the needs of the town with the interests of the country, the wishes of individuals with the broader notion of common good'. We know how to make our cities more compact and attractive; we have the capacity to civilise and link the edgelands with the town and country they confront; we can see a future for a prosperous, more varied countryside to which wildlife may return. Where talented professionals and enterprising local authorities have demonstrated with the help of public-spirited individuals and companies that these objectives can be achieved, the public has responded with enthusiasm.

As climate change begins to affect our lives, so the public becomes more aware that we cannot continue to exploit the physical resources of the earth without thought for the lives of generations to come. Conservation of historic landscapes, local as well as those of national importance, is an integral part of a sustainable future.

Part One

Framework

The historical background
David Cannadine

When Sir Kenneth Clark concluded his television series *Civilization*, he did so with a programme entitled 'Heroic Materialism', in which he explored and explained how science, engineering and technology had transformed the greater European world during the nineteenth and twentieth centuries – in many ways beyond recognition, and in some ways beyond redemption. 'Heroic' described both the achievements of such individuals as Thomas Telford and Isambard Kingdom Brunel, and also the massive scale of the change their endeavours had wrought on the landscape and environment, while 'Materialism' signalled that the creation of wealth, and its unhappy social consequences of poverty, hypocrisy, cruelty and exploitation, was the dominant theme of the times. But it was not just the nature and scope of these developments by which Clark was impressed (and much perturbed), it was also their disruptive speed, as ever more rapid, brutal and irreversible change, at a seemingly exponential rate, and with far-reaching consequences for the world around us, became a permanent feature of modern life. 'Imagine,' he urged his audience, 'an immensely speeded up movie of Manhattan Island during the last hundred years.' 'It would,' he insisted, 'look less like a work of man than like some tremendous natural upheaval. It's godless, it's brutal, it's violent.'

Much the same could be said of Britain, which invented and pioneered 'Heroic Materialism', as a once 'green and pleasant land' was transformed into the first industrial nation and the workshop of the world, duly celebrated in the Great Exhibition of 1851. Across the length and breadth of the country, and long before Manhattan exploded into skyscrapers and imploded into subways, there was 'brutal, godless, violent' change, which damaged, polluted and remade the national landscape. Between 1800 and 1900 the population of England and Wales more than tripled, from slightly fewer than 9 million to roughly 32 million people, most of them crowded into large and growing cities. At the beginning of the period, 25 per cent of English men and women were urban dwellers, while the majority still lived in the country. By the end of the period, 80 per cent resided in London, or in such great cities as Leeds, Liverpool, Manchester and Birmingham, or in the industrialised villages of the Potteries, or in the mining communities of south Wales, or in the hills and dales of south Yorkshire. Nor was the countryside unchanged, as enclosure was followed by high farming, and then by agricultural depression and barbed wire, and as the last great age of country-house building came and went. Moreover, and thanks to the railways, England's towns and countryside were joined up and interconnected as never before – by 1852 there were almost 6,000 miles of track, by 1875 the network had doubled and by 1912 it had almost trebled.

The result of these changes was that by the time Queen Victoria died England had become the most urbanised nation in the world, with the most comprehensive system of railway transport that any country has ever had. Nor was there any let-up during the twentieth century, as the national landscape was unmade and remade, again and again. England's population grew from 32 million to nearly 50 million, cities expanded and suburbanised during the 1920s and 1930s (and almost continually since 1945), and a whole network of new towns was created, from Stevenage to Milton Keynes. By the time of the new millennium, England's urban areas accounted for 90 per cent of the population, economic output and employment. The heavy industries associated with the country's nineteenth-century economic pre-eminence – iron and steel, cotton textiles, shipbuilding, coal mining – have disappeared, and a new service economy has superseded them. The railway system has been na-tionalised, rationalised and privatised, and motorways and new roads have proliferated across the landscape. In 1930 there were one million cars, in 1960 there were 6 million, and in 1990 there were 20 million. Today there are 25 million. As a result, most cities and towns have been redeveloped, with their centres re-planned and their working-class housing cleared. And since the Second World War, farming has become increasingly mecha-nised and large scale.

From one perspective, these changes to England's environ-ment have been all to the good. For they helped bring about, and they were themselves the consequences of, unprecedented improvements in the standards of living of ordinary people. Today in England men and women are better fed, better edu-cated, better housed, better travelled and better paid than at any time in history. They live longer, their working conditions are more congenial, and their social and geographical mobility are greater than ever before. But there has also been a significant downside, as more people, more cities, more industry, more rail-ways, and more cars have exerted immense pressure on the en-vironment, undermining many of the material gains in the standard of living. Nineteenth-century industrial cities were no-torious for the squalor of their working-class housing, and for their polluted atmosphere. Late twentieth-century cities were no less notorious for their poorly planned central areas, their crime-ridden housing estates, their suburban sprawl, and their gridlocked traffic. And the turn-of-the-millennium countryside is in a worse state: retreating from rapidly advancing motorways, bypasses and shopping malls; denuded of its birds, animals, flowers and hedgerows; and facing the inexorable decline in village life and local amenities.

'Heroic materialism', as Kenneth Clark recognised, has been a mixed blessing for our civilisation – for no landscape, be it urban or rural, can survive such disruptions unaltered or un-scathed. Today as in the past, the challenges and dilemmas that face the public, the planners, the pundits and the politicians in looking after our small part of the planet are clear: how to reconcile use with delight, profit with pleasure, development with conservation, the needs of the town with the interests of the country, the wishes of individuals with a broader notion of the common good. But while these challenges are easily specified, they have never been easily addressed or answered. For the unfinished history of conservation in England has an unnervingly equivocal ending. To be sure, many voluntary societies and pressure groups have sought to safeguard the natural and built environment, and to urge the same course of action on successive governments which, since the Second World War, have indeed been increasingly involved in such activities. But the outcome of these endeavours has been at best half-hearted. If we are to make further (and much-needed) progress in protecting and nurturing our environment during the twenty-first century, we would do well to ponder the successes and failures of these earlier efforts, and also the reasons for them.

Pre-history

By 1800, and thanks to a profound shift in sensibilities that had taken place in the preceding three centuries, it no longer seemed self-evidently right for Britons to dominate, exploit and subdue their environment, as experience and reflection increasingly suggested that they should be more concerned to preserve and appreciate the natural world. It was already recognised that towns were becoming polluted, overcrowded and insalubrious, whereas the countryside was clean, quiet and spiritually regenerating; that the fields and forests were in danger of excessive exploitation and cultivation, and should be looked after and taken care of as natural wilderness; and that instead of conquering the environment citizens should protect its species and spaces. These attitudes intensified during the next two centuries, they still inform our contemporary debates about conservation, and most people today would at least pay lip service to them. In the middle of the nineteenth century John Ruskin famously spelt out the parallel case for the conservation of the built environment in *The Seven Lamps of Architecture*. 'It is,' he asserted, 'no question of expediency or feeling whether we shall preserve the buildings of past times or not. *We have no right whatever to touch them.* They are not ours. They belong partly to those who built them, and partly to all generations of mankind who are to follow us.' Here were Edmund Burke's views of society and politics, as the compact between generations dead, living, and yet unborn, transferred to the realms of architecture.

Nevertheless, it was a big step from sensibility to safeguarding, from aspiration to action, for the protection of the environment requires that limits be set to the freedom of individuals or business organisations, especially in regard to the ownership of their property and the size of their profits, in the pursuit of a greater, collective, public good. Before 1914 the very idea that the national landscape might be safeguarded by legislative action would have seemed bewildering and incomprehensible to most people in power. Private property was private property and government interfered with it as little as possible. As Peter Mandler rightly noted in an address to the Royal Historical Society, parliament did not 'legislate for historic preservation and conservation; the British had fewer powers in this area than any other Western European state, both at local and central levels'. Indeed they did. In 1905 the art historian Gerard Baldwin Brown published *The Care of Ancient Monuments*, in which he examined the legislative provision and institutional support (both private and public) for such activities across Europe. His conclusions were sombre. After looking at France and Germany, and also Greece and Denmark, he concluded that 'our official machinery, judged by continental standards, is defective'.

To be sure, parliament did legislate to regulate property rights, notwithstanding the conventional laissez-faire wisdom of the time. But it rarely did so for what we would term conservationist or environmental reasons. Throughout the nineteenth century, parliament regularly passed measures to control the planning, building, sewage arrangements and water supplies of houses, and in the 1870s, 1880s and 1890s, it passed acts in favour of slum-clearance schemes. Yet most of this legislation was permissive rather than mandatory – even when it was applied it could usually be evaded, and the motivation was sanitary rather than environmental. Parliament also intervened in the countryside, as when it passed the Ground Game Act of 1831, which preserved grouse and partridge and pheasant; but it did so only for landlords to shoot, and the population of game-threatening vermin and birds of prey was much reduced in consequence. Similarly, there was legislation concerning the welfare of animals, passed between 1822 and 1835, and further extended thereafter, which outlawed bull running, cockfighting and bear-baiting. But the motive was more to prevent cruelty than to promote conservation.

Only towards the end of the nineteenth century was legislation passed that was more explicitly concerned with conservation and the environment. Some of it was to do with wildlife. In 1869 the Seabirds Protection Act imposed a closed season from 1 May to 1 August for 43 species of bird, and during the next thirty years these provisions were more broadly extended. In 1914 the Grey Seals Protection Act safeguarded mammals for the first time, making it illegal to kill them in their breeding season. There was also one landmark piece of legislation to protect buildings, passed in 1882, when Sir John Lubbock obtained an Act for the Preservation of Ancient Monuments, which was designed to prevent landowners from destroying former settlements on their property, on the grounds that the state had an obligation to protect them for the common good. As a result an inspector was appointed, some monuments were brought under public guardianship, the act was twice renewed and extended in the next 30 years and, in 1908, the Royal Commission on Historical Monuments was created to list and describe buildings and monuments of historical importance throughout Britain for the period before 1700. Concern about the relations between

the city and the country was expressed in the Town Planning
Act of 1909, which enabled local authorities to prepare schemes
for land that was about to be developed, and as such it followed
the views of Raymond Unwin, who constantly urged 'the
drastic and planned improvement of both natural and urban environment'.

But this did not amount to all that much, and as Baldwin
Brown noted, it was certainly unimpressive compared to what
had been accomplished in continental Europe. The legislation
concerning birds and seals was difficult to implement and could
be easily evaded. It had taken Sir John Lubbock more than a
decade to get his ancient monuments measure through (for
which reason it became known as the 'monumentally ancient
bill'); its definitions were loose and it gave inspectors little real
power; and many landowners resented it as an invasion of rights
of property. As for the Town Planning Act, scarcely a handful of
schemes had been proposed before the outbreak of the First
World War. Overall, little was done in terms of preservation and
even less in terms of access. There were many radicals who
looked to parliament to remedy a long and deeply felt griev-
ance, namely that ordinary men and women were unable to
enjoy many of the most beautiful parts of the country because
landowners guarded their property rights zealously. But they did
so in vain. Between 1884 and 1905, James Bryce presented his
Access to Mountains Bill eight times – always unsuccessfully.

With parliament largely indifferent, most of the work for
conservation in the years before the First World War was undertaken by voluntary organisations. In 1865 the Society for the
Preservation of the Commons of London was founded, and
within a decade had saved open spaces at Wimbledon,
Wandsworth and Putney, and also Hampstead Heath. Twelve
years later, William Morris founded the Society for the Protection of Ancient Buildings (SPAB), primarily to safeguard
England's churches from the heavy-handed 'restoration' of Sir
George Gilbert Scott. The Society for the Protection of Birds
began as a ladies' pressure group to oppose the wearing of

plumage (ostrich and game excepted), and it received a royal charter in 1904. The National Trust was founded in 1895, by Octavia Hill, Sir Robert Hunter and Canon Charles Hardwicke Rawnsley, to hold land and to preserve places of historic interest or natural beauty, and in 1907 such land was declared inalienable by act of parliament. In 1897 the Survey of London was established, with support from the London County Council (LCC), partly to provide an inventory of the capital's buildings, but also in the hope that this would make it easier to preserve them at a time when large areas of the city were being demolished and redeveloped.

However, and as with parliamentary legislation, these voluntary societies and endeavours made little impact. Their memberships were small, their influence was tiny, and their public profile was minimal. The Society for the Protection of Ancient Buildings was supported mainly by artists and architects, but apart from influencing ideas and techniques of church restoration, it achieved little. By 1910 it had only 443 members and when the government appointed the Royal Commission on Historical Monuments it deliberately excluded the SPAB's chief spokesman as 'faddist' and 'extremist'. By this time the National Trust had even fewer members and there were real doubts as to whether it would survive. It was regarded by landlords as a threat to private property, and by government as wet and skittish. Even in the era of the Liberal welfare-state reforms, conservation was seen as a threat to laissez faire, and as too backward looking for a powerful and progressive nation. Indeed, the pattern established in the years before 1914 – limited, largely ineffectual legislation, small and crankish pressure groups, and little sympathy or contact between them and government – remained the norm throughout the inter-war years.

Between the wars

At first glance, such lack of progress in the inter-war years seems surprising, for the dominant figure in English politics during this period was Stanley Baldwin, a man who constantly sang the praises of the countryside and the hedgerows. 'To me,' he once observed in his most-quoted words, 'England is the country and the country is England.' He took a genuine delight in being a Worcestershire squire, Mary Webb was his favourite novelist, and among his friends were Sir Edward Grey, the former foreign secretary and a renowned naturalist and birdwatcher, Lord Halifax, a great northern landowner, fox-hunter and believer in 'spiritual values', and John Buchan and G. M. Trevelyan, two famous writers whose attachment to the countryside was well known. But although it was undoubtedly sincere, Baldwin's interest in matters rural was more political than conservational. For him, the countryside spelt decency and respectability, the qualities he was determined to bring back into public life after the corrupt excesses of the Lloyd George coalition; and it spelt neighbourliness and consensus, attitudes he was eager to promote among the broader public in the difficult years of the slump and high unemployment. In short, Baldwin was more interested in using the countryside to the advantage of his politics than he was eager to use his politics to the advantage of the countryside.

Once again, then, we see a legislative record that was distinctly meagre. In the immediate aftermath of the war, the government set up the Forestry Commission, which soon began to plant upland areas with conifers and pines. The primary concern, however, was to grow timber on a sufficient scale to relieve England of the danger of being starved of a strategic resource in the event of another war, and almost from the beginning there were bitter rows between the advocates of commercial forestry and those who wanted to preserve amenity and promote outdoor recreation. In 1932 another Town Planning Act was passed, but it was scarcely more important than its 1909 predecessor. It encouraged local authorities to designate planning schemes for specific areas, but once again its clauses were permissive rather than mandatory, and by the outbreak of the

Second World War virtually nothing had been accomplished under its provisions. The Ribbon Development Act of 1935 was an ineffectual effort to control new housing along arterial roads, and the Access to Mountains Act of 1939 was another legislative non-event. To be sure, it was a measure in the liberal tradition of James Bryce, but it failed to grant any general right of access, left decision-making in the hands of the Ministry of Agriculture, and was soon rendered irrelevant by the outbreak of the Second World War.

This lack of interest may seem all the more remarkable because, as the need for a Ribbon Act suggests, the town and the countryside were undergoing dramatic changes and (as it seemed to some) unprecedented spoliation. In London, redevelopment meant that many of the capital's eighteenth-century architectural glories were demolished, including Lansdowne House, Devonshire House, Nash's stuccoed facades on Regent Street, Soane's Bank of England, Old Waterloo Bridge and the Adelphi. In the countryside building blight seemed to be everywhere, as four million new houses were put up during the interwar years, and as towns and cities expanded on the ground at an unprecedented rate, gobbling up 60,000 acres of rural land a year. The result was a new world of council house estates and white-collar suburbia, of seaside bungalows and holiday homes, as the working and lower middle classes leapfrogged their way out into the country. And they were accompanied by ribbon development, advertising signs, petrol stations, pylons and cinemas. There was also the automobile: between 1924 and 1936 the price of a car fell by 50 per cent and production increased by 500 per cent. Coach parties and day-trippers brought yet more noise and litter to previously tranquil scenes, while the aristocracy, the traditional guardians of the countryside, were in retreat and selling out. In all these ways, prospects seemed bleak.

Inevitably, there were protesting voices, and they came mostly (and predictably) from the upper echelons of society. One of them was Clough Williams-Ellis, who produced two books, *England and the Octopus* (1928), which he wrote himself,

and *Britain and the Beast* (1937), containing essays by such lumi-
naries as J. M. Keynes, E. M. Forster and Patrick Abercrombie,
inveighing against the ruination of the countryside. Another
was G. M. Trevelyan, who became a staunch supporter of the
National Trust and spoke out powerfully against the wounding
of the landscape in *Must England's Beauty Perish?* (1926) and *The
Calls and Claims of Natural Beauty* (1931). With Trevelyan's help,
and under the leadership of John Bailey, R. C. Norman and
Oliver Brett, the scale of the Trust's activity significantly expanded, as both its membership and its holdings increased, from
713 to 6,800 people and from 13,200 to 68,544 acres. In 1926 the
Trust was joined by the Council for the Preservation of Rural
England, which was set up to co-ordinate such activities, and to
bring pressure to bear on the government. And in 1937 the
Georgian Group was inaugurated, to campaign against the demolition of so much of London's architectural heritage, with
active support from Douglas Goldring, Robert Byron, Osbert
Sitwell, John Betjeman, Kenneth Clark and James Lees-Milne.

These were very establishment and well-connected names,
and one of the reasons they achieved relatively little was that
they were too paternal, too exclusive, too elitist for the mass
democracy that the nation had recently become. They were far
more interested in preservation than in access – in keeping the
suburbs and the trippers out of their beloved hills and fields.
'The people's claim,' C. E. M. Joad argued in *Britain and the
Beast*, 'upon the English countryside is paramount.' But, he
went on, 'the people are not as yet ready to take up their claim
without destroying that to which the claim is laid'. And so the
landscape must be kept inviolate 'until such time as they *are*
ready'. But this was not how those in government saw things.
Why should the working classes be denied those rural places
and pleasures that the upper classes had always enjoyed? 'The
countryside,' the preservationists were sternly reminded in an
official memorandum of 1937, 'is not the preserve of the
wealthy and leisured classes.' On the contrary, it was a source of
pride that since the war, 'there has been unparalleled building

development, a development which every government has done its utmost to stimulate, and whose effect has been to create new and better social conditions for a very large number of persons'. From this populist, laissez-faire perspective, 'homes fit for heroes' were much more important than elite spiritual values.

In such a climate of official and popular opinion the conservationists again achieved little. The Council for the Preservation of Rural England put forward ideas concerning green belts to throttle suburbs, strict planning controls to preserve agriculture and scenery, and the appointment of landowners as 'trustees' for the public good. They came to nothing. It did succeed in 1929 in persuading the government to set up a committee to look into the idea of establishing national parks, but its report was lost in the financial and political crisis of 1931, and Baldwin was not prepared to touch it thereafter. The National Trust tried to encourage landowners to join a trusteeship scheme, but they rebuffed it as being pretentious and socialistic. On the whole, both the British public and the British politicians were indifferent (and sometimes hostile) to these preservationist efforts, and so it was scarcely surprising that there was little legislation, and that what there was continued the nineteenth-century practice of being enabling but not mandatory. In 1927 a senior mandarin at the Office of Works noted that Britain had fewer provisions to preserve historic buildings 'than any other country in Europe, with the exception of the Balkan states and Turkey'. And after the Town and Country Planning Act of 1932, which contributed nothing to urban containment or rural development, controls of the built and natural environment remained among the weakest in Europe.

The post-war years

During the Second World War and its immediate aftermath there was dramatic change, not so much because the preservationist lobby triumphed, but because the climate of public and political opinion shifted under the impact of war: towards greater planning of the environment, and also towards wider access to it. This was partly because war wrought havoc with towns and cities: in Coventry and Southampton, 30 per cent of the housing stock was wiped out, and a similar proportion of LCC council houses was destroyed or harmed. Most country houses were requisitioned, many suffered lasting damage, and it seemed unlikely that their owners would be willing or able to return at the end of the war. But it was also because Labour dominated the home front in Churchill's coalition and won the election in 1945, and so set in train a series of inquiries, which served as the blueprint for its management of the environment in its years of power, not so much as individual property, but rather as collective heritage. Among them were the Barlow Commission (1940) on the distribution of the industrial population, the Scott Committee (1942) on planning in rural areas, the Dower Committee (1945) on national parks, the Hobhouse Committee (1947) on national parks and access, and the Gowers Committee (1948–50) on houses of outstanding historic importance or architectural interest.

From these reports and inquiries came a flood of legislation that set the framework for conservation for a generation. The Town and Country Planning Act of 1944 authorised the listing of individual buildings of historic significance, going far beyond the narrow category of scheduled monuments. In 1946 Hugh Dalton set up a £50 million National Land Fund, 'to buy some of the best of our still unspoiled open country and stretches of coast, to be preserved forever, not for the enjoyment of a few private landowners, but as a playground and a national possession for all our people'. A further Town and Country Planning Act of 1947 introduced the concept of comprehensive development plans across the whole urban and rural landscape. In 1949 the National Parks and Access to the Countryside Act was passed: it gave provision for the designation of National Parks and Areas of Outstanding National Beauty, and a Nature Conservancy was set up to establish a network of National Nature

Reserves and Sites of Special Scientific Interest. And in 1953, in the aftermath of Gowers, the Historic Buildings and Ancient Monuments Act established the Historic Buildings Council, which distributed repair and maintenance grants to the owners of great but threatened mansions in return for guaranteed public access.

These wide-ranging and interlocking measures were intended to usher in a whole brave new world of active conservation on behalf of everyone, in marked contrast to the limited legislation and paternal preservationism of the early twentieth century and the inter-war years. Looking after the environment, both urban and rural, was now the responsibility of government, and during the next two decades, regardless of whether Labour or the Conservatives were in power, that responsibility further increased. In 1951 the Rivers (Prevention of Pollution) Act was passed and in 1956 the Clean Air Act, which within ten years transformed the environment of industrial towns and cities. In 1957 Duncan Sandys established the Civic Trust, which co-ordinated the activities of local amenity societies working to preserve the built-up landscape, and which became an urban equivalent of the Council for the Preservation of Rural England. Ten years later Sandys piloted through the Civic Amenities Act, which made it possible to establish conservation areas in cities, towns and villages. A Town and Country Planning Act of 1968 tightened up the listed-building procedure, the Countryside Act of the same year gave the Countryside Commissioners a general responsibility for safeguarding the landscape, and the Conservation of Wild Creatures and Wild Plants Act of 1975 established a framework for the protection of all types of wildlife.

Yet these well-meaning attempts to safeguard and plan the national environment seem to have met at best with limited success. In the cities there was destruction and redevelopment on a massive and wholly unprecedented scale. Between 1955 and 1975, 1.3 million people were rehoused, moving from slum dwellings into high-rise flats, which were hugely subsidised, but were soon disliked as unfriendly and unwelcoming council houses in the sky. As a result, neighbourhoods were demolished, communities destroyed, streets obliterated: in 1970 the GLC acknowledged that 67,000 houses that it had bulldozed could and should have been renovated. At the same time, historic city centres were wiped out, and replaced with soulless office blocks, multi-storey car parks, and impenetrable ring roads. For this was the era of property developers like Jack Cotton and Charles Clore, and of large-scale building contractors such as Bovis, Wimpey and Taylor Woodrow. Many famous and historic buildings were razed to the ground, among them the Birmingham Public Library, the Midland Institute and the Doric Arch at Euston Station. Despite public pressure, much of it organised by the recently founded Victorian Society, the prime minister, Harold Macmillan, authorised the destruction of the Euston Arch in 1961, on the grounds that conservation was only a minority interest, and that such obsession with the past would sap the nation's vitality. Small wonder that in 1975 Colin Amery and Dan Cruickshank published a book on the vandalising of cities entitled *The Rape of Britain*.

At the same time, the destruction of country houses also reached unprecedented levels. The 1950s and early 1960s were peak years for demolitions: at least 10 per cent of the national stock disappeared completely, including such illustrious piles as Eaton Hall and Panshanger. Some great houses were much reduced in size, like Bowood and Woburn, others were taken over by the National Trust under its country houses scheme, and many more were sold off for institutional use as schools, colleges, hospitals and nursing homes. Some finance was initially made available from the National Land Fund to pass a dozen or so properties to the National Trust, but by the early 1960s the Fund was essentially moribund. In such a climate of devastation, John Harris and Marcus Binney mounted an exhibition at the Victoria and Albert Museum in 1974 on 'The Destruction of the Country House' and soon after, a new pressure-group was formed called SAVE Britain's Heritage. It was far more publicity conscious than the inter-war organisations had been, but all the campaigning and lobbying in the world could not prevent the sale of Mentmore Towers in 1977 and the dispersal of its fabulous collections – as a result of which the by-now defunct National Land Fund was revived as the National Heritage Memorial Fund.

Nor did the countryside fare any better, for as in the towns, the protective legislation was simply not strong enough. Because of the overriding need to produce food and timber domestically, in the aftermath of the war, agriculture and forestry were both exempted from the development control of the Town and Country Planning Acts. Although ten national parks had been established by 1960, there were none in Scotland, they were never nationally owned, they were inadequately administered and overseen, and it proved impossible to prevent economic development and intensive farming. Copper mining was allowed in Snowdonia, an oil terminal was established on the Pembrokeshire coast, and the early-warning station was built at Fylingdales on the North Yorkshire Moors. Despite 3,000 Sites of Special Scientific Interest having been designated by 1975, this proved to be little more than a label stuck on the ground, and which the farmer could disregard at will. In the same way, the Countryside Commissioners had been given inadequate powers to exert any serious form of control, and the Conservation of Wild Creatures and Wild Plants Act listed too few species, many of which had already disappeared altogether. In the cities, it seemed, the planners had been given too much power; in the countryside, by contrast, they did not seem to have enough.

In many ways the real threat to the rural environment in these post-war years came not so much from continuing urban encroachment (although there was, indeed, plenty of that) but from newly mechanised farming, which soon began to wreak havoc on the flora and fauna of the countryside. In the aftermath of war it was thought essential that Britain should continue to be self-sufficient in food, and the Agriculture Act of

1947 brought about guaranteed markets and assured prices for farmers, which lasted for a generation. This was accompanied by the revolution in agricultural technology, in which horses were replaced by tractors, and by the unprecedented use of fertilisers and pesticides. In 1940 there were 1,100 tractor-mounted sprayers in England and Wales; by 1981 there were 74,000. In 1944 there were 63 products approved for use by farmers as pesticides; by 1976 there were 819. Increased mechanisation made for bigger fields, which meant the destruction of meadowlands, grasslands, downlands and hedgerows. Between 1947 and 1985 hedgerow length fell by 25 per cent and this, combined with the growing use of chemicals, had a devastating effect on flowers, trees, animals and birds. By 1964 peregrines were at 44 per cent of their pre-war numbers, and many other species suffered similar fates. In 1977 the Nature Conservancy Council produced a report emphasising the damaging effects that new-style agriculture was having on the biodiversity of the environment, and in 1980 Marion Shoard published an angry, wounded book lamenting *The Theft of the Countryside*.

Underlying all this was a growing dissatisfaction on what might, during this period, be termed the two evolving sides of the great conservation divide. For those who cared about the natural and the built environment, it seemed that the initially welcomed increase in government involvement had gone terribly wrong. In *The Making of the English Landscape*, W. G. Hoskins regretted that the countryside he had known and loved was vanishing, and was being replaced by the planners, the politicians and the vandals with a new, horrible, 'barbaric England' of 'the arterial by-pass, treeless and stinking of diesel oil, murderous with lorries'. For their part, those in government, regardless of political affiliation, were no less annoyed that the growing cult of conservation was being used as a brake on what they saw as much-needed progress and improvement. In a speech in 1962, Lord Hailsham, himself no philistine, urged that 'all the really artistically healthy societies of the world have been marked with a supreme artistic self-confidence'. 'Mattocks and sticks of dynamite' were their watchword, 'to clear away the rubble of the past, often of exquisite beauty, and to make way for the beauties of the future.' In a later Tory government, Peter Walker deplored the public's sentimental tendency 'to retain all that exists and to oppose all that is new', while for Labour, Anthony Crosland took the populist line that the conservationists were 'hostile to growth and indifferent to the needs of ordinary people'.

Where we are now

During the last twenty years it has become much more difficult for opponents of conservation to make such accusations of class bias and elitism, as the urge to protect the environment has simultaneously become unprecedentedly popular in its appeal, and unprecedentedly global in its reach. Hence Greenpeace and Friends of the Earth, hence the growing concern about global warming, hence the Prince of Wales's advocacy of organic farming. Such activities encompass a broad spectrum of political opinion, and an equally broad swathe of society. In England the 1980s saw an unparalleled rise in the number of people joining conservationist organisations. For example, in 1945, the Royal Society for the Protection of Birds had a membership of only 5,900; in 1980 the figure was 320,000 and by 1997 it was one million. The National Trust shows the same trend: 200,000 members in 1970; one million in 1980; two million in 1990; and 2.7 million in 2001. No political party in Britain today can remotely compete with such large-scale popular support and involvement. Even Margaret Thatcher recognised the direction in which the wind was now blowing. 'We are all no more than life tenants of our heritage,' she observed in a speech to the Royal Fine Arts Commission in 1989, expressing views that would have done credit to Ruskin, 'and we have a moral duty to pass it on in as good a condition as that in which we received it'.

As a result, the conservation climate warmed up during the

1980s, with those in favour, and those previously against, seeming to reach some sort of mutual understanding and accommodation. For its part, the government became more sympathetic. The National Heritage Act, passed in 1983, replaced the Historic Buildings Council with English Heritage, 'an independent body devoted to the conservation and presentation of England's inheritance of ancient monuments and historic buildings'. As Secretary of State for the Environment, Michael Heseltine was a powerful advocate of comprehensive listing, and in the mid-1980s the National Heritage Memorial Fund gave £25 million to 'save' the three great houses of Kedleston, Nostell and Weston, and to prevent any repetition of the Mentmore debacle. And after the planning excesses that had blighted so many inner cities, ancient and modern, during the 1960s and 1970s, there was now a growing recognition that the conservation of a more human and appealing urban environment was the precondition for renewal, not the enemy of it. Accordingly, when Marcus Binney published *Our Vanishing Heritage* in 1984, he noted that conservationists were now more willing and able to 'stand up and speak out for endangered buildings', and his book was not so much a lament for what had been lost as an account of country houses, public buildings, churches, terraces, town houses and factories that had been successfully saved and often put to new and imaginative uses.

Underlying these changes was a deeper shift in mood against the ethos of planning and control that had been the guiding (though not always effective) force in conservation since the Labour government of 1945. One indication of this was the intervention of the Prince of Wales in public debates. Between them, he claimed, the planners and architects had done more damage to England's towns and cities than the Luftwaffe, and in *A Vision of Britain* (1989) he set out an alternative view of urban and rural landscapes in which a sense of the past, of place, and of human scale was given higher priority than utopian, ahistorical and anti-human blueprints. This was the reassertion of the patrician aesthetic of Clough Williams-Ellis and his friends. But the

planners were equally assailed from the other side of the political spectrum by the free-market populism of Margaret Thatcher.
For she regarded them with the same loathing with which she
regarded bishops, academics and civil servants: they were elitist,
self-perpetuating, irresponsible, and indifferent to the needs and
views of ordinary men and women. It was 'the planners', she
told the Conservative Party Conference in 1987, who had 'cut
the heart out of our cities'. Accordingly, the construction of
inner-urban motorways was brought to an end, no more mu-
nicipal flats were built, many existing blocks were demolished,
and council houses were sold off.

The result of these changes in outlook and action was that
between 1979 and 1997 there was an increasingly relaxed attitude towards planning. In some ways this improved the prospects for conservation, but in other ways the effect was
decidedly adverse. By the late 1980s the Secretary of State for
the Environment, Nicholas Ridley, was bullying local authorities into granting planning permission for new developments on
greenfield sites at the edge of many towns (while objecting to
similar practices in his own neighbourhood). This in turn led to
a new and largely unregulated suburbanisation of the countryside, concentrated around shopping malls and housing estates. In
1986 Britain contained 432 superstores; in 1996 there were
1,034. Since 1986 retail giant Sainsbury's has added 206 new
supermarkets: 10 per cent in town centres; 23 per cent in
suburbs; 67 per cent on outside sites. At the same time, the major
house-building contractors covered the adjacent fields with
homes in standardised and repetitive style that are insensitive to
the local built environment and traditional materials, and thus
are unappealingly reminiscent of Lego. 'No continental European country,' the historian Richard Rodger claims, 'has cloned
private housing on such a scale in the last quarter of a century.'

In all these ways, planning deregulation has been more the
enemy than the handmaid of conservation: both in the town
and in the country, the increasingly unfettered free market,
driven by populist impulses and laissez-faire ideology, has

blighted the environment in recent times to an even greater extent than during the inter-war years. As stores have moved out from city centres to city peripheries, the variety and vibrancy of downtown life have been much diminished, as smaller shops close down and many buildings become derelict, thereby compounding the mistakes made by the planners in earlier decades. More out-of-town shopping means more out-of-town car journeys: since 1980, road traffic has increased by two-thirds and is still growing. As a result, Britain's newly sprawling towns and cities are more polluted and more gridlocked than ever before. Indeed, according to Lord Rogers's Urban Task Force, which reported in 1999, the quality of urban life in Britain has fallen far behind that enjoyed by much of Europe. Meanwhile, the advance of the urban frontier into the country seems both inexorable and irresistible. At present, England is losing 27,000 acres (nearly 11,000 hectares) annually to urban development and if this trend continues until 2050, the total built-up area will be twice as great as it is today.

Nor are these the only problems that beset the conservation of the countryside. Today, agriculture contributes only 1 per cent to gross domestic product, and employs a mere 1 per cent of the labour force. Yet these employees are responsible for the stewardship of 80 per cent of the nation's land surface. In lowland regions, and notwithstanding restrictions on the use of chemicals, over-subsidised farming continues to ruin the landscape. Hedgerows still disappear at an alarming rate, the number of starlings halved between 1972 and 1997, and many other species have similarly declined. Meanwhile, and as the recent outbreak of foot-and-mouth disease has searingly shown, much upland farming, in the West Country, the Lake District and the Pennines, is scarcely viable economically: if this section of the countryside is to be maintained and conserved new ways will probably have to be found to finance it. Small wonder there are many who cry that 'the countryside' is in 'crisis'. It certainly is from a conservation point of view. 'The landscape,' Oliver Rackham wrote in 1986, in words that have become even more

pertinent since, 'is a record of our roots and the growth of civilisation.' 'Almost every rural change since 1945,' he goes on, ' has
extended what is already commonplace at the expense of what
is wonderful or rare or has meaning.' This is not exactly encouraging.

What hope?

It should by now be clear that the history of conservation in
twentieth-century England has, at best, an equivocal outcome,
for it is in many ways a story of lost causes, disappointed hopes,
and failed endeavours, as a result of which both the urban and
rural landscape, have suffered irreversible damage. On the other
hand, we should not lose sight of the fact that in conservation,
as in everything, the best is often the enemy of the good. For
notwithstanding the traumatic changes to which the environment has been subjected during the last two hundred years, the
fact remains that many parts of England, both city and country,
are still extraordinarily beautiful, and even today, after a century
of unprecedented urban encroachment, nearly 90 per cent of
the nation's surface area is still classified as being rural. We should
also remember that constraint, which is the essential key to conservation, is an exceptionally difficult political (and philosophical) issue. As Simon Schama has observed, there are 'profound
problems' that beset any democracy seeking both 'to repair environmental abuse and to preserve liberty'. Today, the most effective way to improve our environment at a stroke would be to
reduce pressure on it by banning second children, second cars,
and second homes. In a few years this would transform our cities
and our countryside for the better. But such draconian measures
are not possible in a democracy; and nor should they be.

By definition, democracy is a cumbersome instrument for
dealing with complicated issues, and that is certainly so in the
case of conservation. 'The planners' are blamed: sometimes for
having too much power, sometimes for having insufficient. 'The

market' enables individuals to satisfy their own desires (and who are any of us to gainsay *that*?), yet it takes no account of the general good. Inevitably, the result is a bewildering array of half-thought-out and half-completed initiatives by a no-less bewildering array of voluntary societies, government departments and (increasingly) EU directives and global imperatives. As Oliver Rackham notes, 'historic landscapes and historic buildings are similar in many ways, and both should have the same kind of legal protection'. But, he adds, 'the case for conservation is weakened by lack of co-ordination between those concerned with scenery, wildlife, antiquities and freedom'. This is not only true in terms of voluntary societies, but also in terms of government, as exemplified by the seemingly endless reconfiguring (and renaming) of those many departments (transport, agriculture, defence, environment and heritage) that have a legitimate concern with conservation. The results are those shuffles and scuffles, the compromises and bargains of politics, which so enrage the most zealous friends of the earth, for whom the death of nature is imminent, and the stark alternatives are those of extinction and (just maybe) redemption.

Put less apocalyptically, this means that in the future as in the past, there will always be a conflict between the use of the environment to sustain our material way of life and the enjoyment of the environment to nourish our other (but no less important?) needs. What are the prospects for the countryside, when subsidised and mechanised farming has wrought such damage to our lowland landscape, when upland agriculture is in economic crisis, and when the advance of the urban fringe continues unhalted? How will our country and our cities cope with the projected 4.4 million new homes which will, apparently, be needed by 2021, and which will be the equivalent of 25 additional towns the size of Milton Keynes? And how can new life be brought back into the centre of our cities, and the planning mistakes and excesses of the 1960s and 1970s be remedied, at a time when road traffic is set to increase by another third in the next twenty years? During the past decade, a succession of governments has produced a succession of white papers on Environmental Strategy, Urban Life, and Rural Conditions. But so far their impact has been negligible. The preservation of our environment requires serious political engagement, committed public involvement, and the achievement of consent and consensus. But as the history of the twentieth century's efforts at conservation shows, these are not easily obtained.

In any case, can we afford to be thus non-apocalyptic? As the rainforests disappear, as tigers are threatened with extinction, as Venice seems set to sink beneath the waves, and as global warming intensifies, the answer may well be that we cannot. For the conservation of our national environment is but a small part of the larger enterprise of conserving our global environment. And if it is difficult to reach an agreed and viable policy within *one* nation, how will an agreed and viable policy ever be evolved between *all* nations? Faced with such pressing and intractable problems, all the historian can do is to point out the successes and failures of the past, and offer some tentative explanations as to why some things happened (for good or ill) and others didn't (ditto). Like all historical accounts of contemporary issues, this essay is less a recipe for action than an invitation to reflection, and as such it is meant as a contribution to self-knowledge rather than a strategy for ecological rescue. But for all that, a better understanding of how we arrived at our current local predicaments might, just possibly, help to clear our heads about which directions to take in the future. For if the history of conservation efforts in twentieth-century England proves anything, it is that clear heads are at least as important (and rarely as much in evidence) as good intentions.

The changing countryside: the effects of climate change on the landscape

Crispin Tickell

It is a common delusion that the landscape was as it is and will be. Our language and literature are dotted with such phrases as 'old as the hills' and 'mountains eternal'. We take comfort from the apparent stability of our surroundings against the ups and downs of daily life and the ephemeral nature of life itself. Our lives are indeed short, and our instinctive view of the world is equally so. Yet as Heraclitus said 2,500 years ago, there is nothing permanent except change.

In the last two hundred years our horizons have been painfully stretched. Space and time now look very different. We live on a small planet around a middle-sized star in the arm of a galaxy containing 200 billion stars, itself no more than one among billions of comparable galaxies reaching to the limits of telescopic inquiry. As for time, the earth is around 4.6 billion years old, and throughout its history its surface shape has changed immeasurably. A great heat engine beneath our feet drives the plates on the surface above and below each other like ill-fitting bits of mosaic. So it will continue until the earth is eventually burnt up by the sun. Every now and again the earth is struck and sometimes transformed by impacts from space. The thin atmosphere of gases that surrounds it, like bloom on an apple, is in constant movement as it conveys heat from one part to another, and reacts to the changing relationship of the earth to the sun.

No wonder the only constant in climate is its inconstancy. Climate and landscape are shaped by the ever-changing distribution of land, sea and living organisms; and in turn land, sea and living organisms help to determine climate and landscape. On a fairly recent timescale the collision of India with the Asian land mass pushed up the Himalayan mountains: it diverted air and ocean currents, created the Asian monsoons, increased rock weathering (thereby enabling living organisms to draw carbon out of the atmosphere), and indirectly changed temperature and weather patterns worldwide. More recently still, the closure of the sea passage between North and South America had comparable results. Indeed, it may have helped precipitate the beginning of the Pleistocene ice ages, which began less than two million years ago.

At present we live in a warm patch between glacial fluctuations, some big, some small, within a broad 100,000-year rhythm. During glacial periods sea levels sink and land bridges appear, for example between Alaska and Siberia, between the islands of Southeast Asia, between Ireland and Britain, and between Britain and France. Vast ice sheets cover parts of both hemispheres, affecting climate deep into the tropics. During warm patches, as some 120,000 years ago, the ice shrinks, sea levels rise, and the world we know, or something like it with a subtly different distribution of living organisms, reappears. The timing of the next glaciation is a matter of debate. Indeed, it is arguable that glacial rather than warm periods are the norm, and that our present warm patch is an anomaly.

The British landscape well illustrates the oscillations. In the last warm period, what is now Trafalgar Square was a semi-tropical swamp with exotic plants and home to the hippopotamus. A mere 20,000 years ago it was part of a windswept tundra inhabited by reindeer, woolly rhinoceros and mammoth. When the present warm period began, around 11,000 years ago, it

formed part of the delta system of the Thames in which the site
of Westminster Abbey was an island covered in thorns. Traces of
the last ice age were everywhere. Most of Britain was dense
forest and took long to clear. Even in the last thousand years
there have been marked changes. From the ninth to the twelfth
century Britain was probably warmer than it is today, and grapes
were grown; a little ice age followed in which from time to time
the Thames was frozen over, and fairs were held on the ice. The
present warming began only a century or so ago, with a rise of
global surface temperature of between 0.6°C and 0.8°C, which
had multiplying effects on landscape and the living creatures
within it.

So the landscape is about as inconstant as climate, although
there is a time lag between them. In the normal way of things
climate will continue to fluctuate, landscape will continue to
respond, and ecosystems will continue to adapt themselves to
change, with greater or lesser success and many casualties on the
way. But now there is the new element, unknown in previous
history: the impact of one animal species – our own – on the
global environment.

In the past, human activities, like those of beavers and other
creatures, have had a local and sometimes regional effect. For
example, the burning of grasslands in Africa, deforestation in
Mesopotamia and the Indus valley, the heavy farming of the
southern shores of the Mediterranean by the Romans, and the
agricultural practices of the pre-Columbian peoples of Mexico,
all changed the landscape, and in most cases changed and sometimes damaged the resource base on which society was built.
This was as true of Britain as anywhere else. Sometimes the
landscape can recover, as in parts of Central America. More
often the wounds are long and deep. Flying over the sand and
camel grass that now surrounds the ancient city of Petra, it is
hard to imagine how the area could have supported a population of around 60,000 people. But the stone walls of the old
fields are still there, as they are in Sicily and other areas ruined
by intensive exploitation by our species.

An interesting question is whether human activity changed not only the landscape but also in doing so the prevailing local climate. The answer is that it probably did, even in temperate Britain. As has already been seen in Amazonia, deforestation on any scale changes patterns of rainfall as well as the capacity of soils to retain moisture. In the same way, cities create heat islands with strong effects on local weather. Any driver can test this going in or out of towns of any size. Once processes of this kind are set in train, it is impossible to say whether previous conditions will be restored. In some cases, with time, they will be. But climate is a global phenomenon, and is subject to so many interconnections that any equilibrium is itself ephemeral. Once changed it can be changed for ever, and the landscape with it.

The problem today is of a different order of magnitude. Since the industrial revolution began, about 250 years ago, humans have not only been changing the landscape at an accelerating rate but their activities have also changed the chemistry of the atmosphere. It is notoriously difficult to distinguish natural from man-made processes, but there is a growing consensus, expressed in successive reports from the Intergovernmental Panel on Climate Change (which brings together the vast majority of the world's experts on the subject), that the human contribution is now having a significant if not decisive effect. Indeed, in its guidance to policymakers, Working Group I (Science) of the Intergovernmental Panel concluded that 'in the light of new evidence and taking into account the remaining uncertainties, most of the observed warming over the last 50 years is likely to have been due to the [human-induced] increase in greenhouse gas concentrations'.

There is a clear relationship between the concentration of greenhouse gases and the surface temperature and thus character of the earth. These gases are mainly carbon dioxide (the most important greenhouse gas after water vapour), methane, and nitrous oxide. The increase in atmospheric carbon dioxide, arising from combustion of fossil fuels and such other factors as changes in land use (in particular deforestation), has been over 30 per cent since 1750, and is now at its highest for almost half a million years. Over the same period the increase in methane, more than half of it from human activity, has been over 150 per cent, and increase in nitrous oxide, about a third of it from human activity, has been 17 per cent. Atmospheric concentration of all these gases is increasing, and in the case of carbon dioxide, the rate of increase is accelerating, although many of the effects have yet to be seen. At the same time the capacity of land and sea to absorb carbon dioxide is likely to diminish.

This is not the place to go into current negotiations, arising from the Framework Convention on Climate Change of 1992, to limit the emission of greenhouse gases. For present purposes it is enough to say that even in the unlikely event that the reductions foreseen in the Kyoto Protocol were to take place, they would make little difference to the current build-up of greenhouse gases worldwide. Kyoto was never more than a start. It will be centuries before current and future concentrations of greenhouse gases disperse. The question is to decide what level of concentration to go for in the perhaps vain hope of achieving some eventual stability in the future. In the meantime, the immediate practical issue is how best to adapt to the certain prospect of change.

How warm would a warmer world be? According to Working Group I of the Intergovernmental Panel, 'the globally averaged surface temperature is projected to increase by 1.4°C to 5.8°C over the period 1990 to 2100'. This is a considerable increase on the 1.0°C to 3.5°C rise suggested in its previous report of November 1995. It covers a wide range of local variations. It would, for example, be different in different parts of Britain. But overall 'the projected rate of warming is much larger than the observed changes during the twentieth century, and is very likely to be without precedent during at least the last ten thousand years'.

What would be the characteristics of the landscape of a warmer world? Here the uncertainties multiply. Efforts have

been made by Working Group II (Impacts) of the Intergovernmental Panel to assess possible impacts by continent, but the
results are inevitably sketchy for such a relatively small area as
the British Isles. However interpreted, they suggest a different
world and a correspondingly different distribution of human activity, as people and the living organisms on which they depend
try to adapt to change. Such change includes new patterns of
rainfall and drought, as well as more extreme events, and rising
sea levels.

There are two jokers in the pack. The first is that the melting
of arctic ice could alter the ocean circulation system in the
north Atlantic. At present, the flow of warm salty water north-
east across the Atlantic is roughly equal to that of a hundred
river Amazons, and results in an enormous transfer of heat.
Without it winter temperatures in Western Europe would fall
dramatically: Liverpool or Dublin could have the temperature of
Spitzbergen or the Hudson Bay. Some 12,000 years ago the
warming that ended the last ice age led to just such changes in
the Atlantic; and Western Europe and to some extent other parts
of the northern hemisphere had a mini ice age of over a thousand years until the currents – known as the Atlantic Conveyor
– resumed their previous track. Paradoxically, global warming
could thus lead to at least regional cooling.

The second joker would be equally unwelcome. In the long
past there has been strong warming as well as strong cooling of
the earth's surface. Methane – molecule for molecule – is a
much stronger greenhouse gas than carbon dioxide, and has
been released from beneath the oceans and the frozen tundra
when a certain threshold of warming has been passed. The result
has been a sudden rise in global surface temperature and for a
while the establishment of a hothouse earth. Something like
this happened, for reasons not yet clear, at the end of the Palaeo-
cene epoch some 55 million years ago. It could certainly happen
again.

Drastic changes of this kind are unlikely, but they are not
science fiction. As knowledge of the past has increased, so has

understanding of the volatility of climate and the fragility of any equilibrium. As Professor Fairbanks of the Lamont-Doherty Earth Observatory at Columbia University once said of climate change:

> the best analogy is to the cogs in a watch, turning at different speeds, some of them directly connected, others not. People are distracted by the predictions of gentle, long-term global warming, a few degrees spread over centuries. But small perturbations of climate can lead to large consequences, and they do not necessarily have to be gradual . . .

The history of the periodic Niño phenomenon in the Pacific well illustrates this point. The last Niño, in 1997–8, had effects all over the world. Now Niños seem to be getting stronger and more disruptive.

There is an additional factor that profoundly affects landscapes. Through their activities humans are at present destroying other living organisms at a rate that can be compared in its effects with the five great natural extinctions of the last 440 million years. After the impact of an extraterrestrial object some 65 million years ago, when the dinosaur family finally disappeared, it took between 5 and 10 million years for the earth's living system – the biosphere – fully to recover. We have little idea of what the results of the current extinction event will be, and how it will affect such predatory species as our own. In the processes of change climate is one of the main drivers, and our descendants, if there be such, will inhabit landscapes of greatly different composition and covering.

Even in many of our own lifetimes there have been perceptible changes in Britain. It would be easy to shrug them off as mere variations in weather. After all, nothing drastic in our temperate island has occurred. Yet the trends are clear. The warmth of the 1930s gave way to the cooler 1950s and 1960s, before the steady rise in temperature witnessed since then. Over the last century the landscape changed in many ways, some in response to climatic variation, but most as a result of such human activity as intensive farming, including hedge clearing, and urban sprawl, with consequent redistribution of ecosystems.

Whether quick or slow and in whatever form, the present global trend is towards warming, with wide regional and local variations. On that assumption, however provisional, every country can begin to work out possible effects, and to some extent prepare for them. Within Europe, according to Working Group II of the Intergovernmental Panel, the south and the arctic are the most vulnerable to change. In the south there could be less rainfall in the summer and more in winter, and river flood hazards could increase. Half of the alpine glaciers and large permafrost areas could disappear and there could be an upward and northward shift of ecosystems with loss of important habitats.

Predictions for Britain made by the government's Climate Impacts Programme tell broadly the same story, but in more detail. The models suggest that there will be continued warming, especially in winter, and the overall rate of warming will range from 0.1°C to 0.3°C a decade during this century. There would be a southeast to northwest gradient across Britain with the south-east warming more rapidly than the north-west. There would be more precipitation, particularly in winter, with a tendency to summer droughts in the south. There would be more extreme events, such as storms, droughts and gales, and a continued sea level rise of between 2 and 9 centimetres a decade with increased hazard of storm surges and high tides. Here the problem is confounded by another factor. Since the last ice age much of southern and eastern Britain has been sinking, and northern and western Britain have been rising in relation to sea level. London has a particular problem with rising groundwater.

These changes, if they were to happen, and still more the combination of them, would obviously have widespread impact on our society and its economy. They would also affect the landscape in more ways than can now be easily imagined. A visitor in 2100 would find a map of the physical geography of Britain significantly different from what it is today. Some coast-lines would have had bites taken out of them, giving East Anglia and southeast England a different shape. A higher Thames barrage would be protecting London, and fortified banks would stretch on each side of the barrage towards the sea. Coastal de-fences would have been redeployed in many cases. With peri-odic flooding and erosion of steep-sided valleys, familiar landscapes, particularly in Scotland, Wales and southwest England, would be subtly different. Parts might look more like the northern Mediterranean and others more like southern Scandinavia. Overall we would see changes in the human impact, ranging from some redeployment of towns and ports, new lakes and reservoirs, shifted centres of industry and power generation, corresponding changes in the network of roads and railways, and different uses of agricultural land and forestry.

Less obvious would be changes in ecosystems and the distri-bution of wildlife, from animals, birds and plants to fish, insects and micro-organisms, with all the implications for human health and welfare that these changes bring. The ecology of such spe-cialised habitats as the Cairngorms could be drastically altered. Tree growth in Scotland, Wales and northwest England would be accelerated, while in the south-east loss of salt marsh, mud-flats and lowland heaths would affect birds and other popula-tions. The distribution of insects and micro-organisms would also be different – they are normally quick to respond to change. In the warm period in the early Middle Ages, malarial mosquitoes reached as far north as Stockholm, and could do so again.

For thousands of years the British landscape has been partly shaped by human activity, and natural features have been over-lain by human ones. Climate change would thus be the agent for the transformation of one partly artificial landscape for another. It is a nice question how much we could – or would – wish to keep the landscape broadly as it now is. East Anglia is a case in point. Already there is slow erosion of the coastline, and inland some areas are reverting to the conditions that existed before the Fens were drained. Resistance to rising sea levels and sinking land would be in vain, as well as prohibitively costly in most cases, and priorities for action would have to be established. The same goes for other parts of the country where planning for change has already begun. Those early to grasp the implications could reap enormous advantage. The effects will reach far and wide through the economy, from the distribution of population to styles of building and use of natural materials.

Perhaps what is new is the likely speed of change. For most of us in most of history, climate change and its effects on landscape are like a slow film, telling a story in which we have to pinch ourselves to stay awake. Now the story goes fast forward. It is not a story limited to climate. The myriad changes we are making to our surroundings produce combinations of change that make us more giddy than sleepy. As when visiting cherished places of childhood we find that they are both the same but disturbingly different. They remind us how vulnerable we are as small creatures among the swirling gases on the earth's surface, which is all our living space.

Population pressures

John I. Clarke

There can be little doubt that Britain's relatively small land mass is heavily populated, and that it is experiencing growing population pressures, manifested by overcrowding, traffic jams, pollution, poverty and continuous expansion of the built environment. Those pressures are felt unevenly across the country, but mostly within a relatively small part of its total area, particularly within and around its urban concentrations, along its major transport links and where both are insidiously encroaching upon the countryside.

At the beginning of the new millennium, the United Kingdom contained nearly 60 million people. It had grown by just under 2 million a decade since the middle of the twentieth century. This was largely by natural increase – the number of births exceeding the number of deaths. That volume of growth is likely to continue for the next two decades, but rather more by net inward migration than by natural increase. Yet for decades many have said that our country has already too many people and that any more would pose additional burdens upon various aspects of society and the environment. For example, nearly thirty years ago the geographer and MP Edwin Brooks in *This Crowded Kingdom* (1973) was concerned about population growth in Britain leading to the destruction of natural resources, the spread of built-up areas, the urbanisation of the countryside, the 'tidal erosion of village culture' and the emergence of a single large megalopolis. Those concerns are still commonly expressed, but both the population size and its overall living standards continue to rise, with consequent increases in population pressures upon the environment.

The main demographic impact of population on the landscape and countryside will not be merely through the addition of a few extra million people, but more through a series of other population processes including:

- the growing contrasts between the inhabited and uninhabited areas of Britain;
- the continuation of the long-term demographic decline of the north and west of Britain and the increasing concentration in the south and east of England;
- the dominance of the rather nebulous and ill-defined urbanised axial belt running diagonally across England from the south-east through London and north-north-westwards to Manchester, Merseyside, Lancashire and West Yorkshire;
- the localised effects of the recent surge in net inward migration;
- the effects of changes in population structure, notably the inexorable increase in the numbers of old people and the decline of the traditional nuclear family system, on the rapidly growing numbers of households and dwellings; and
- the massive multiplication in human mobility, facilitated by rising living standards and the booming popularity of the motorcar, enabling major urban decentralisation, especially of retailing.

In other words, the impact of populations upon the British countryside does not result only from the addition of large numbers of people, but also from the changing characteristics of those people. Populations are extremely dynamic, and cannot be easily controlled or directed.

Population size and growth

Although the surface area of the United Kingdom (242,910 square kilometres) is moderate by international standards, in 2001 it contained an estimated 59.95 million people (unfortunately the 2001 census results are not yet available), 58.24 million or over 97 per cent of whom lived within Great Britain. This put the UK in twentieth place in population size of the 190 independent countries in the world, though its ranking is continually falling (it was sixteenth in 1990), not only because the number of independent countries increases fitfully and was boosted recently by the fragmentation of the former federal republics of the USSR and Yugoslavia, but also because its annual population growth rate of about 0.4 per cent at the end of the twentieth century was much slower than the world average rate (1.2 per cent), particularly when compared against the average rate of the developing countries (1.7 per cent).

As is well known, the UK is one of the most densely populated of the major countries in the world, with a mean density in 2001 of about 247 people per square kilometre, which is two and a half times the population density of France, a country with a similar total population. Some smaller countries are more densely peopled (e.g. Belgium, Mauritius, Puerto Rico, Singapore), but among those with more than 50 million inhabitants it is only exceeded by Bangladesh, India, Japan, Philippines and Vietnam. Obviously, crude population densities of countries are not very meaningful, as all countries have extremely varied population distributions that are generally becoming more uneven as the process of urbanisation concentrates populations in particular areas. In the UK, as in most countries, extra population does not fill up the empty spaces of uninhabited areas, but tends to congregate within the already densely populated ones. Its uneven population lies somewhere within a very wide range of uneven population distributions, such as those of Algeria, Argentina, Australia, Canada and Egypt, whose populations are extremely localised on tiny parts of their territories, and those of

Bangladesh, Belgium and Germany, whose populations are moderately evenly spaced. However, population pressure is not realised by crude overall densities, but in the containment of a large and growing population with changing structural characteristics, rising living standards and increasing mobility. In the UK, more than in any of the other populous countries of Europe (e.g. France, Germany, Italy, Poland, Spain, Russia and Ukraine), living space is at a premium, although Germany runs it very close.

During the nineteenth century and the first half of the twentieth century the population of the UK grew from 11 million to about 50 million, largely by a net natural increase of births over deaths, and this occurred despite long periods of substantial net outward migration to other continents, especially to the United States and parts of the Empire. Since the mid-twentieth century, natural increase has oscillated through fluctuating but generally declining numbers of births and remarkably constant numbers of deaths, but over the last two decades the gap between births and deaths (or net natural change) has diminished greatly to a little over 100,000 a year, less than a third of its average during the 1960s (see Figure 1 and Table 1). In the 1990s natural increase averaged only 0.19 per cent per annum.

So at the beginning of this twenty-first century, natural increase of population in the UK has slowed down markedly to a relatively stable low level of roughly 100,000 a year. Like many of the populations of West European countries, there is no great potential for natural increase at present or in the foreseeable future, and there is even the possibility of long-term natural decrease (as seen so strikingly at the moment in many East European countries, particularly the Russian Federation), unless there is an unexpected resurgence in fertility to offset the growing proportion of older people.

Overall population growth during the next two decades will depend greatly on a continuation of the net flow of inward migration that assumed growing importance in the 1980s and particularly during the late 1990s (see Table 1). Whereas the UK

Table 1 **Annual population change in the United Kingdom, 1951–2021** (in thousands)

	Population at start of period	Net natural change	——— Annual averages ——— Net inward migration	Overall change
Mid-year estimates				
1951–61	50,287	246	12	258
1961–71	52,807	324	–14	310
1971–81	55,928	69	–27	42
1981–91	56,357	103	43	146
1991–9	57,814	107	104	211
Mid-year projections (in 1998)				
1999–2001	59,501	88	140	228
2001–11	59,954	87	95	182
2011–21	61,773	92	95	187

Source: National Statistics, *Social Trends* 31, 2001, p. 33.

had experienced a net loss of population from international migration during the first three decades of the twentieth century and later during the 1960s and 1970s, from 1983 onwards it experienced a net gain. Net inward migration during the period 1991–9 averaged more than 100,000 per annum, nearly matching the amount of natural increase (Table 1), but this was mainly due to the surge in net inflows of 178,000 in 1998 and 182,000 in 1999. The main reason for the increase in net inward migration during the late 1990s and early 2000s was the number of asylum seekers who were accepted for settlement, though in most years only about 20 per cent of asylum applicants were recognised as refugees or granted leave to remain in the UK. The net inflow means that an increasing proportion of the total population have community backgrounds outside the UK; the Prakesh Report[1] estimated them in 1998 at 5.75 million people, or just over a tenth of the population of Britain at the time.

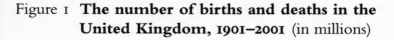

Figure 1 **The number of births and deaths in the United Kingdom, 1901–2001** (in millions)

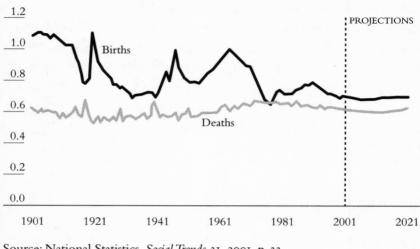

Source: National Statistics, *Social Trends* 31, 2001, p. 33.

Net inward migration is expected to continue, although annually it fluctuates so strongly and the whole issue of asylum seekers is so politically volatile that it poses particular problems for the always difficult task of making population projections, thus considerably reducing their reliability. Much will depend upon government attitudes and controls over the movements of migrants. The 1998-based population projections by the Office for National Statistics (ONS) assume that during the early decades of the new millennium, net inward migration will continue to be very significant, averaging about 95,000 a year and probably exceeding natural increase as a factor in overall population growth (Table 1).

According to those population projections, the UK population will grow by 0.3 per cent per annum until 2021, when it will contain 63.64 million inhabitants, about 3.7 million more than in 2001. In short, it is expected to grow by a little less than 2 million a decade, roughly the absolute rate at which it grew during the second half of the twentieth century, albeit then

largely by natural increase. Longer-term projections suggest that
the population of the UK will reach a peak of 65 million in the
year 2036, when it will start to decline. However, such projections are unreliable, given that it is assumed that future growth
will be predominantly by net inward migration. No doubt unforeseen events will continue to falsify projections of linear
trends. In fact it has been suggested that: 'the one certainty of
making population projections is that, due to the inherent un-
predictability of demographic behaviour, they will turn out to
be wrong as a forecast of future demographic events or population structure'.[2]

Growing population concentration

A remarkable feature of population growth in the UK during
the last two centuries has been the increasing proportion living
within England and the declining proportions living within
Scotland, Wales and Northern Ireland, a process that is expected
to persist at least during the first two decades of the twenty-first
century. Between 1851 and 2001, the population of England
grew from 16.8 to 50.2 million, from 75.3 per cent to 83.7 per
cent of the total UK population, and projections suggest that it
will increase by a further 3.5 million to reach 53.7 million in
2021, which equals 84.4 per cent of the total UK population. In
contrast, the combined total in 2001 of the much smaller populations of Scotland, Wales and Northern Ireland was less than 10
million and they had diminished to only 8.5 per cent, 4.9 per
cent and 2.9 per cent respectively of the UK population. Their
relative demographic decline is expected to continue; by 2021 it
is projected that the percentage of the UK population living in
Scotland, Wales or Northern Ireland will decline from 16.3 per
cent to 15.7 per cent, and that population will still number less
than 10 million (Table 2).

This trend reflects a relative demographic decline in the
north and west of Britain and an overall growth in the south and

Table 2 **Populations of the countries of the United Kingdom, 1851–2021**

	1851	1901	1951	2001 est.	2021 est.
	Millions				
Great Britain	20.82	37.00	48.85	58.24	61.82
England	16.76	30.52	41.16	50.19	53.72
Wales	1.16	2.02	2.60	2.95	3.05
Scotland	2.89	4.47	5.10	5.11	5.06
Northern Ireland	1.44	1.24	1.37	1.71	1.82
United Kingdom	22.26	38.24	50.23	59.95	63.64
	Percentages				
Great Britain	93.5	96.8	97.3	97.1	97.1
England	75.3	79.8	81.9	83.7	84.4
Wales	5.2	5.3	5.2	4.9	4.8
Scotland	13.0	11.7	10.2	8.5	8.0
Northern Ireland	6.5	3.2	2.7	2.9	2.9
United Kingdom	100.0	100.0	100.0	100.0	100.0

Source: National Statistics, *Social Trends*. 31, 2001, p. 30.

east. This may also be expressed in broad terms as a contrast between Highland Britain in the west and Lowland Britain in the east, usually considered as being on either side of a line from the mouth of the Tees to the mouth of the Exe, a divide first identified at the beginning of the twentieth century by Sir Halford Mackinder in his classic volume *Britain and the British Seas* (1902), and still of some utility. The more sparsely populated Highland Britain, where vast expanses of uninhabited mountains and moorlands abound and where localised population agglomerations were often associated with the traditional industries of coal mining, iron and steel production, shipbuilding, textile and chemical industries, has been characterised by long-term decline in population. Being located entirely within Highland Britain, the population distributions of Scotland and

Wales are therefore much more localised than those of England. Most of England lies within the much more uniformly populated Lowland Britain, where empty spaces are rarer, towns and cities more regularly distributed and new light manufacturing and service industries more attracted by proximity to mainland Europe, and which has experienced consistent population growth. It is of course only a rough dualism rather than a sharp dichotomy, as by this definition Highland Britain contains populous lowland regions, notably parts of the Central Lowlands of Scotland and South Wales, that bear more demographic affinity with Lowland Britain and represent the population growth areas of their respective countries.

Growing population concentration in the south and east and population stability or decline in the north and west have been a fairly persistent process in Britain since the end of the First World War, and are exemplified by regional population changes during the 1990s (see Table 3). While the populations of the London, South East, East, South West and East Midlands regions grew at rates ranging from 3.9 per cent to 5.7 per cent during 1991–9, those of Wales, Yorkshire and the Humber, and the West Midlands grew by only 1.3 per cent to 1.6 per cent, Scotland was almost stationary and the populations of the North East and the North West regions experienced slight declines. So by the end of the twentieth century a quarter of the population of the whole of Britain was living in two English regions, London and the South East, a proportion that is likely to grow. In the short-term, sub-national ONS population projections of 1998, London is identified as the region with the greatest projected increase of population, at 7.1 per cent between 1998 and 2008, followed by the South East (6.5 per cent), East (6.1 per cent), South West (5.8 per cent) and East Midlands (4.4 per cent). In contrast, the North West and North East regions are expected to experience small declines of 0.6 per cent. In short, the future pattern of regional population changes looks like more of the same.

During the 1990s the London, West Midlands, Yorkshire and Humber regions gained much more from natural increase than

Table 3 **Estimated regional population change in Great Britain, 1991–9** (in thousands)

Region	1991	1999	Change	% change	Natural change	Migration & other changes
North East	2,603	2,581	-21	-0.8	4	-25
North West	6,885	6,881	-5	-0.1	52	-56
Yorkshire & Humber	4,983	5,047	64	1.3	58	6
East Midlands	4,035	4,191	156	3.9	51	105
West Midlands	5,265	5,336	70	1.3	93	-23
East	5,150	5,419	269	5.2	97	172
London	6,890	7,285	395	5.7	316	80
South East	7,679	8,078	399	5.2	110	289
South West	4,718	4,936	218	4.6	-7	225
England	**48,208**	**49,753**	**1,545**	**3.2**	**773**	**772**
Scotland	5,107	5,119	12	0.2	3	9
Wales	2,892	2,937	46	1.6	6	39

Source: J. Horsfield, 'Population review of 1999: England and Wales', *Population Trends* 102, Winter 2000, pp. 5–12.

from net migration, whereas the South East, South West, East and East Midlands regions gained much more from migration (Table 3). A good deal of that migration initiates from London, which acts as a reservoir for internal migrants to other regions, especially for retirement, as well as a clearing house for many of the international migrants that it receives. In general, the south-eastern part of Britain, including London, is regarded as an escalator region[3] that creams off the natural increase of other regions, and of Scotland and Wales, by providing opportunities and attractions to a disproportionate number of educated and skilled young adult migrants, internal and international. At the same time, it loses a disproportionate number of middle-aged professionals and older retirees to other regions, particularly in the south of England.

Net inter-regional migration, however, gives no indication

of the volume of regional in-migration and of out-migration. In any one year, a vast number of inter-regional movements cancel themselves out, so that the net changes in population distribution are small. This is a characteristic feature of many of the more developed countries, and has been termed 'dynamic equilibrium' in relation to internal migration in Australia.[4] It is evidenced, for example, in the small regional balances and limited overall impact of the two and half million inter-regional movements that took place within the UK during 1999. During that year, the London region experienced the greatest regional net outward movement of 65,000 internal migrants, so supplementing the populations of neighbouring regions, but this loss was more than counterbalanced by a net inflow of international migrants.

Uninhabited and inhabited areas

Any detailed map of Britain's current population distribution or density demonstrates clearly its unevenness, with large empty areas contrasting with the many metropolitan concentrations (see Figure 2). At least a third of the land area of Britain is uninhabited and another sixth is very sparsely inhabited at less than 25 inhabitants per square kilometre. Consequently, the inhabited area occupies only half the total land but contains 98–9 per cent of the total population. It is difficult to define precisely the threshold between inhabited and uninhabited areas, as it depends largely upon the size of unit areas considered. Does one person living in 1,000 square kilometres constitute an inhabited area? Hardly. In fact, the broad pattern of uninhabited areas closely reflects the harsher environments and is therefore remarkably stable, never having been densely peopled. On the other hand, these areas are expanding little by little because of rural depopulation of the sparsely peopled hill-farming areas that have suffered from acute problems associated with BSE and foot-and-mouth disease, and from the consequent decline in tourism.

Figure 2 **The 1991 distribution of population across
Great Britain by smallest census areas**

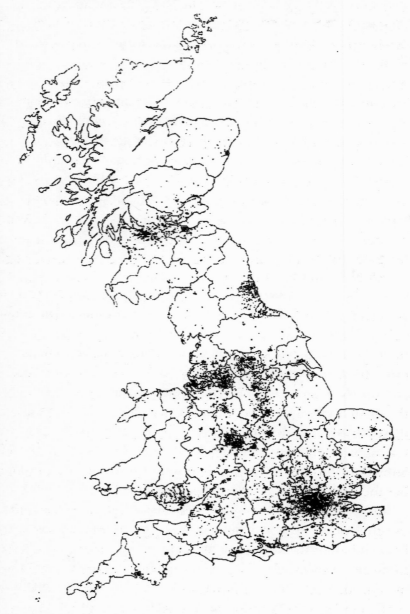

Source: Dorling (1999), p. 26. Unfortunately, the 2001 census data are not
      yet available.

The largest uninhabited and sparsely peopled areas have long coincided with the mountains and moorlands of Highland Britain, greatly influenced by altitude and relief: the Scottish Highlands and Islands, the upland districts of central Scotland, the Southern Uplands of Scotland, the Lake District, the Pennines, the North Yorkshire Moors, the Welsh Massif and the massifs of South West England (Bodmin Moor, Dartmoor and Exmoor). To them must be added much smaller uninhabited and sparsely inhabited areas in Lowland Britain: the summits of many of the limestone and chalk scarplands of England, the New Forest in Hampshire, the Brecklands of East Anglia, Dungeness, the Essex marshlands and the North Kent coasts. The fine details of these empty areas became especially evident in the intricate patterns of population distribution elucidated by the kilometre-grid-square data that became available for the first time after the 1971 census of Britain.[5]

All these uninhabited and sparsely inhabited areas, many of which have been designated Areas of Outstanding Natural Beauty and National Parks, have limited and restricted residential locations, but as the car-owning population of Britain continues to increase they are visited in normal circumstances ever more frequently by tourists and second-home owners. Leisure and tourism have become the dominant economy of many of the small towns and villages situated within and around some of these largely uninhabited areas, as seen, for example, in the Cairngorms, the Lake District, the Pennine Dales, Exmoor and Snowdonia, so that their transient populations greatly outnumber their permanent residential populations.

Of course, the rurality of these more remote areas differs greatly from that of the rural areas found within the more inhabited half of Britain. Similarly, there are considerable contrasts in the proportions, densities and characteristics of the rural populations of Highland and Lowland Britain. In Highland Britain rural populations constitute over a quarter, and in places over a third, of the total population over large swathes of the Highlands and Islands, Dumfries and Galloway, the Scottish Borders,

Northumberland, Cumbria, the Northern Pennines, as well as much of Wales, the Welsh Borders and the South West. In contrast, in most of Lowland Britain that level of rural population is only found in North Yorkshire, Lincolnshire and East Anglia; elsewhere it is frequently less than 10 per cent or even 5 per cent.[6] As most of the rural areas of Lowland Britain are within commuter range of a major city, they are much less isolated and in closer contact with the towns and cities that dominate the pattern of population distribution.

Urban concentration in the axial belt

Although the terms urban and rural are in constant use, their demographic distinction has never been clear. Definitions of urban have varied greatly between countries, being based on different criteria (administrative status, built-up areas, economic functions and population size and density), none of which satisfactorily clarifies a dualism that everyone knows about but which is ever more arbitrary as town and country intermingle in so many ways.

Whichever definition is used, we can still say that some nine-tenths of the total population of Britain live in urban areas, the proportion changing little in recent decades and differing little between Scotland, England and Wales. Perhaps surprisingly to the great majority of us who live within urban areas and have an urban-orientated view of the landscape and countryside, they occupy a tiny fraction of the total land area; the Office of National Statistics found that in 1991 urban areas occupied only 6.0 per cent of the total land area of Britain, and although the percentage was much higher in England — where populations are highly localised — than in either Wales (3.1 per cent) or Scotland (2.8 per cent), it was still only 8.3 per cent.[7] The inevitable increase over the next few years is unlikely to be dramatic. Of course, urban areas are very unevenly distributed within the three countries, and there is an enormous range in the percent-

Expanding urban areas

Having sketched the macro-geographic distribution and redis-tribution of population, it is appropriate to look at some of the more important micro-geographic processes affecting detailed patterns of population distribution and to look at how these en-croach upon the countryside.

One major micro-geographic movement since the 1960s is the way in which the prolonged process of rural-urban migra-tion has been largely replaced by a complex of inter-urban, intra-urban and urban-rural movements. Large cities and conurbations experienced urban deconcentration and decen-tralisation to the suburbs, to smaller towns and villages and to surrounding rural areas, a process sometimes called counter-urbanisation, and which was greatly facilitated by the mercurial growth in personal transport. It meant a dispersal of population and this had major impacts upon the countryside, especially in the more populated parts of Britain. However, suburbanisation around most towns and cities was constrained to a greater or lesser extent by green belt planning controls and by rising house prices, leading to out-migration to smaller towns and increased commuting distances. Excluding London, during the 1990s the six metropolitan counties either declined in population or re-mained roughly stable, while many medium and small-sized towns and villages continued to grow, especially in the more economically favoured areas south and east of a line from the Severn to the Humber. Consequently, although the percentage of the total population living in urban areas has remained roughly the same at just below 90 per cent, the total number of urban areas in Britain increased from 2,231 in 1981 to 2,307 in 1991, when the census recorded a particularly large increase in the number of smaller urban areas with less than 10,000 inhabit-ants in the shire counties. In other words, the range of city sizes in Britain is becoming slightly less hierarchical.

Inevitably, the rate of change has varied over time and in dif-ferent parts of the country, and during the 1990s there was a

slowing down of counter-urbanisation as some of the disadvantages of population dispersal became apparent (e.g. remoteness
from major retail outlets, increased journey-to-work times and
fuel costs). In addition, different groups of people were moving
in different directions. People moving out from inner cities were
generally older and were usually replaced by younger people
either living alone or with families, and often of immigrant
origin. Consequently, apart from that of London, the populations of metropolitan areas are projected to decline only slightly
– by about 1 per cent – between 1998 and 2008, although naturally there are regional variations: while West Yorkshire is expected to increase by about 2 per cent, Merseyside is expected
to decline by 4 per cent.

London is exceptional. Its previous demographic decline was
arrested and turned around during the 1980s and 1990s, especially through net inward migration from abroad. The Prakesh
Report estimated that in 1998 about 35 per cent of the population of Greater London had community backgrounds outside
Britain and that 22 per cent had backgrounds in Africa, the
Caribbean or Asia. That turnaround has accelerated, particularly
from the late 1990s with the rise in the numbers of international
migrants. In 2001, it was estimated that in Greater London there
were about 60,000 asylum seekers and 200,000 refugees, and
that about 85 per cent of all asylum seekers entering Britain go
to London. It is not surprising that the population of London is
projected to increase by 7 per cent between 1998 and 2008 –
inner London by over 8 per cent and outer London by over 6
per cent.

Changing population structures and households

Changing population distributions are not merely a matter of
population numbers, but also reflect changing population structures and their impact on the numbers of households and the
stock of housing. To the effects of recent net inward migration

into Britain and the localisation of ethnic minority groups in London and other cities within the axial belt must be added the even more significant effects of the progressive ageing of the population.

Although at the beginning of the twentieth century only 5 per cent of the population of the UK was aged 65 and over, by the end of the century improvements in the mortality rates of older people and lower fertility rates had raised the level to nearly 16 per cent, and it is still rising. Rates vary between men and women. The gender gap is currently wide, as only 13 per cent of men are 65 or more compared with 18 per cent of women, an increasing number of whom are living alone widowed, divorced or separated. That gap will almost certainly grow; 1998-based ONS projections suggest that by 2026 the respective percentages will be 15 and 22. They also suggest that the median age of the UK population will rise from 36.9 years in 1998 to 41.8 years by 2021, when the number of people of pensionable age is projected to increase from 10.7 million to 12.2 million.[11] By 2008 the number of pensioners is expected to exceed the number of children, and the age structure will become even less pyramidal and more rectangular in shape. In short, there is a huge greying of the population that is common to all European countries and that has manifold effects on the number and composition of households, the location and type of housing, and the level of integration or segregation of the aged population.

Another important demographic factor that will continue to affect the numbers and compositions of households, and consequently the development of housing, is the considerable fluctuation in the size of birth cohorts since the Second World War. Added to this are social factors, associated particularly with the progressive decline in the size and the stability of the nuclear family: the falling number of married-couple households along with the rising numbers of lone-parent and cohabiting households and of people living alone, especially among the elderly.

As a result of demographic and social processes, phenomenal

changes took place in household numbers and size in Britain during the twentieth century. While the population increased by more than half, the number of households more than tripled from just under 8 million to over 24 million and average household size almost halved from 4.6 to 2.4 persons. One of the most remarkable changes, and one which is not always fully appreciated, was the rise in the number of one-person households from just over 2 million in 1961 to nearly 8 million in 2000, when they accounted for almost three in ten of all households (Table 4). Moreover, by 2000 less than three in ten of all households were so-called 'traditional' family households containing a couple with dependent or non-dependent children, compared with nearly one in two in 1961. Similar numbers were recorded for households containing couples with no children.

Household sizes and compositions show regional, urban/ rural and ethnic differences. For example, in England one-person and lone-parent households are most common (and elderly people least common) in the London region, but Scotland also has a high percentage of one-person households. In addition, South Asian households tend to be much larger than those of other ethnic groups, while nearly half of households headed by a black person were lone-parent families. Unfortunately, segregation of ethnic minority households is tending to increase, at a time when integration is politically far more desirable.

If current trends continue, it has been calculated that between 1996 and 2021 there is likely to be a growth of 3.8 million households in England alone, of which over three-quarters will result from changes in the numbers of adults and the overall age structure,[12] with two-thirds of them located in the South East, London, East and South West regions. During that period, average household size is projected to continue to fall, from 2.40 to 2.15 persons; one-person households are expected to form 71 per cent of the net increase in household numbers, while a projected decline of 1 million married-couple households will be more than replaced by an increase of more than 1.3 million cohabiting households.

Table 4 Changes in household size in Britain, 1961–2000

	1961	2000
	— Percentages —	
One person	14	29
Two people	30	35
Three people	23	16
Four people	18	14
Five people	9	5
Six or more people	7	2
Couple with no children	26	29
Couple with children	48	29
Lone parent with children	6	9
Multi-family households	3	1
All households (millions)	16.3	23.9
Average household size (persons)	3.1	2.4

Source: National Statistics, *Social Trends* 31, 2001, p. 42.

It should also be noted that virtually all of the projected growth in household numbers will be among the over-45 age groups, including the post-war 'baby boomers', but with the inevitable rider that these projections are particularly sensitive to future fluctuations in international migration.[13] The picture will be roughly the same in Wales and Scotland, despite their projected limited growth or decline in population. Between 1996 and 2021 the number of households in Wales is projected to increase by 0.2 million to 1.3 million and the average household size to decline from 2.5 to 2.2. In Scotland, the number of households is expected to increase by 12 per cent from 2.17 million in 1998 to 2.43 million in 2012. One-person households may increase by 234,000 from 32 per cent to 38 per cent and average household size fall from 2.3 to 2.0.

The housing picture

Major household changes have led to massive changes in the stock of dwellings in Britain. During the second half of the twentieth century total stock rose by about 78 per cent from 13.8 million dwellings in 1951 to 24.6 million in 2000. No wonder there is concern about the continual growth of the built environment. Almost two-fifths (38 per cent) of the dwelling stock has been built since 1965, when for the first time the number of separate dwellings in Britain began to exceed the number of households. Two-thirds of our purpose-built flats or maisonettes have also been built since then, and there has been a further shift in favour of detached housing. Although the growth in the total number of dwellings slowed down some-what during the 1990s, it was still increasing by about 180,000 a year at the end of the century. In addition, the percentage of dwellings that are owner-occupied more than doubled in the second half of the twentieth century from 29.6 per cent in 1951 to 67.8 per cent in 2000, a huge transformation.

The housing picture has always been far from uniform across the country. There are considerable differences in the type of household accommodation (detached, semi-detached, terraced, flats) between urban and rural areas in Britain and between the various regions, with evident differences in the impact on the landscape. For example, remote rural areas of Highland Britain have vastly higher-than-average proportions of detached houses, three-quarters of households in the North East of England live in either semi-detached or terraced housing, English urban areas have high proportions of older terraced housing, and over a third of households in Scotland live in flats, and far more in urban areas.

Recent regional increases in the housing stock reflect the changes in the numbers of households. During the last two decades of the twentieth century, household numbers grew by more than 20 per cent in the South West, East and South East of England, twice the rate of growth witnessed in the North East and North West. Such regional trends are likely to continue. In

London and many other metropolitan areas, the rising number of younger one-person households places special demands on housing provision, particularly in the form of flats and housing conversions in inner-city areas. Rapidly rising university populations have led to conversions of older terraces on a massive scale, so that some parts of cities (e.g. Heaton in Newcastle and Lenton in Nottingham) have become densely peopled student dormitories. At the same time, increasing numbers of older one-person households, especially for the widowed and divorced, under-occupy other terraced and detached housing. Life-cycle changes mean that households and housing never match.

Inner-city housing, therefore, is changing constantly while further expansions are simultaneously taking place at the fringe. Between 1994 and 1998, 13,165 hectares changed to urban use in England alone, 44 per cent of which had been previously in rural use. All regions were affected but once again changes were more marked in the south than the north. Although the government's target is for 60 per cent of additional housing (new or conversions of existing buildings) to be built at higher densities on previously developed 'brownfield' sites (land that is no longer needed for its original use) by 2008, for many years to come there will still be very many tens of thousands of detached and semi-detached houses constructed at lower densities on previously undeveloped greenfield sites around the fringes of most of our towns and cities, particularly in the south of England.

Housing estates in the suburbs and on the urban fringes are designed to meet the aspirations of the growing 'middle classes' of Britain who wish to escape some of the problems of inner cities in order to enjoy some of the benefits of proximity to rurality. Suburban estates are more popular in highly urbanised Britain than in almost any other country of the European Union, and they continue to grow. Unfortunately, the current difficulties being experienced by the agriculture industry will probably open the door still wider to such expansions.

Hypermobility and the honeypots of the urban fringe

Urban expansion is not just a matter of residential moves to the urban fringe, but the decentralisation there of many urban functions. A multitude of out-of-town supermarkets, shopping centres, industrial estates, science parks, garden centres and sporting stadia has sprouted around towns and cities large and small all over Britain, serving the populations of very large areas. Many are on greenfield sites, replacing more centrally located corner shops, co-operatives, department stores, smoke-stacked factories and traditional football grounds hemmed in by ter- raced rows. At the top of the retailing range are about a dozen 'megamalls', such as the Metrocentre in Gateshead, Meadow- field in Sheffield, the Trafford Centre in Manchester and Brae- head in Glasgow, which are advertised as regional shopping centres, with vast car parks to attract clients from long distances.

These new honeypots of the urban fringe, like many other honeypots in the heart of the countryside (e.g. National Trust and English Heritage properties and 'natural' beauty spots), have evolved with the growing popularity of the motorcar, which enables daily travel of people to work, shop and leisure from ever-widening radii of urban and rural settlements, thus redu- cing the necessity for residential mobility of the population. They are both a cause and an effect of the increasing hypermo- bility of British society, with average daily journeys getting steadily longer. It has been estimated that the average daily travel by Britons multiplied 6 times in 50 years, from about 5 miles a day in 1950 to 30 miles a day by the end of the century.[14] Cen- suses do not provide data of such travel, but all the evidence sug- gests that it will carry on growing, despite attempts to stem the flows, and that it will continue to have a major influence on the impact of the British population on the countryside.

Growing pressures on the countryside

In the past there have been many calls for zero population growth in Britain or for a smaller population, in order to sustain the quality of life of the people and the ecological well-being of the country, and to avoid environmental degradation. As long ago as the early 1970s it was argued that the population should fall even to below 40 million,[15] and more recently others have suggested that a long-term optimum population for the United Kingdom would be in the region of 30 million, about half its current population. Such proposals have usually assumed significantly reduced numbers of children, but have failed to take into account the substantial immigration currently taking place in Britain. To achieve such targets within the next hundred years or so would necessitate not only very much lower fertility than at present (about 1.7 children per woman) but also a reversal of the current net inflow of international migrants. At the beginning of the new millennium all the signs are in the opposite direction, with a small but steady natural increase and a similar amount of net inward migration leading to a continuation of the present fairly gradual increase of about 2 million people a decade.

Most of that increase will be concentrated in Lowland Britain, particularly in large and small urban areas in the south and east of England and, to a lesser extent, other parts of the axial belt. It is there where the pressures of an expanding built environment will be felt most of all, in inner and outer cities that are the main foci of employment and in many smaller towns and villages within commuting range, especially of London.

Sustainable development – or faster, further and more?

David Banister

The planning system has provided an enormously successful means to control development in Britain, and this is reflected in the landscape we enjoy today. However, there are new pressures for housing and other development, which have been augmented by the increase in mobility brought about by the use of the car. There is also anxiety about the continued exploitation of non-renewable resources and about environmental quality. These factors have resulted in the emergence of a new framework for planning and transport that embraces a much wider range of activities than the traditional physical concerns of planners. It addresses the balance between the desire for economic growth and the equally important issues of the environmental, social and spatial implications of that growth. In other words, it is concerned with the achievement of sustainable development.

A recent government paper has described sustainable development as being about

> ensuring a better quality of life for everyone, now and for generations to come. It is concerned with achieving economic growth, in the form of higher living standards, while protecting and where possible enhancing the environment – not just for its own sake, but because a damaged environment will sooner or later hold back

economic growth and lower the quality of life.
Sustainable development is equally concerned with
making sure that these economic and environmental
benefits are available to everyone, not just to a privileged
few.[1]

The challenge for planners is to bring such a concept to
reality. This requires a fundamental rethinking of their role as
guardians of the public interest in order that they may become
much more actively involved in the future of town and country.
Planning must move from being a control function to being an
initiator of change. Apart from a fundamentally different approach to development, this also calls for a rethink of the necessary skills, powers, resources and organisation. This is where the
difference between the rhetoric and the reality becomes apparent, and where good intentions have not yet been effectively
implemented.

Planning, transport and sustainable development

Although it has been argued that planning is central to sustainable development, the concept itself is far wider. The UK Sustainable Development Commission has focused the debate on
six main areas:[2]

1   access, including transport, mobility and the
    planning/land-use dimension;
2   improving eco-efficiency with a particular focus on
    production eco-efficiency;
3   health, including decoupling it from sickness
    treatment, and the role of lifestyle and consumption
    choices;
4   food production and consumption, with a focus on
    agriculture and rural land use, including
    environmental management of the rural landscape;

5 participation, leadership, citizenship and
   government;
6 joined-up government and the evidence-based
   approach.

Planning impinges on several of these issues, but we will
focus on the first, together with some reference to the other
factors where necessary, principally 5 and 6.

Underlying much of the debate on sustainable development
is the importance of accessibility, in other words the ability to
reach a range of services and facilities easily. This simple
concept, at the heart of much planning and transport thinking,
aims at keeping travel distances as short as possible. One of the
fundamental axioms of transport planning is that people do not
like travelling (though we will see later that this is not always
the case), and that they only travel because the benefits received at the destination (e.g. work, education) more than outweigh the costs (i.e. time and money) of getting there. An
important corollary of this axiom is that if journey distances
are short, then people are more likely to use 'green modes' of
transport − walking, cycling or public transport. As journey
lengths increase so does the likelihood that the car will be
used. The second corollary is that the solution depends less on
transport than on land use and development. If decisions are
made to locate new housing, shopping, employment, recreational and other facilities in close proximity to each other,
then journey lengths should be reduced, as people have the
opportunity to do things locally. This will maximise accessibility and, by integrating transport and planning, substantially increase the chances of sustainable development. An accessibility
strategy such as this also meets the environmental and equity
concerns of sustainable development, as people with no car
available can reach local facilities easily.

However, it is recognised that location alone will not necessarily persuade people to use those facilities. For example, they
may choose to travel further to the supermarket, rather than use

the local shop, to take advantage of a wider range of goods. Of course they have a right to do so, but debates over participation and citizenship must be undertaken so that individuals are made aware of the consequences of their choices for sustainable development. We are all part of the problem and so we must also be part of the solution. Although increased transport services can be provided in order to give more choice, the most important determinants are where the facilities and services are located. This is where land-use planners and transport planners must work closely together.

The principles of accessibility are clear, as is their importance in contributing towards sustainable development. But when we look at what is actually happening in practice the evidence is not encouraging. Each person in Great Britain makes about 1,000 journeys a year, for five main types of activity – work, education, shopping, social and leisure. Although the total number of trips has remained constant since 1985–6, the total distance has increased by nearly 28 per cent – we are now travelling much further to carry out the same number of activities than we were fifteen years ago.

The picture is further endorsed by an examination of the modes of transport used. The dominance of the car has increased and now accounts for 61 per cent of all journeys and 79 per cent of all distance travelled. Although walking and cycling account for 29 per cent of journeys (down from 37 per cent only twelve years earlier), these modes are mainly for short local trips and so make up only a small proportion of total distance travelled (about 3.5 per cent). Public transport has also fallen from 10 per cent of journeys to 7.6 per cent, and rail accounts for only 1.5 per cent of journeys, but 5 per cent of distance.[3]

The growth in road traffic has not been evenly spread across all roads. Over the period 1989–99, motorway traffic has grown by 42 per cent in Britain, whilst traffic on built-up major roads with a speed limit of 40 mph or less has remained stable. On other major roads in the suburbs and rural areas, growth has been 21 per cent over the decade, as people disperse to lower-density locations and choose to travel by car to neighbouring towns and the countryside.

The message is clear, namely that the car has increased its dominance at the expense of all other modes of transport. Even the healthy green modes have suffered. All the indicators show that we are in fact moving away from rather than towards sustainable development. Transport's contribution to sustainable development over the recent past has been a negative one. If we take other indicators, such as energy use and levels of some major emissions, the same conclusion is reached. Transport now consumes 81 per cent of all petroleum products in the United Kingdom and is a significant contributor to carbon dioxide (26 per cent) and other emissions.[4] Technology has a major role to play here through add-on technology, principally the catalytic converter and particulate traps, and through more efficient engine design. However, the technology is yet to be discovered that will reduce the level of CO_2 emissions, transport's main contribution to global warming. If carbon-based fuels continue to be used, then CO_2 will be produced in ever-increasing quantities from transport, unless innovation radically improves engine efficiency and accelerated replacement of old vehicles takes place.

Two reasons for increasing car dependency and journey lengths have been the closure of many services and facilities, particularly in rural areas, and consolidation and specialisation within outlets. Often there are substantial cost savings to the provider of the service (e.g. the Health Service or the local education authority) resulting from the economies of scale gained from closing local facilities. But these costs are simply passed on to the users of the service who have to travel further to get to the facility. Only a social audit on proposals to close a facility would establish the overall welfare gain or loss resulting from the closure. That audit should form part of the decision as to whether a particular facility should stay open or not and a subsidy could be paid to the provider.

Similarly, a social audit of any new development would assess not only the benefits to the local economy and users of the new facility, but also the additional costs in terms of travel distances and impact on the town centre. It would extend the notion of the sequential test currently being used for new shops and residential development. This is a test placed on developers to demonstrate that there is no suitable alternative location for their proposal within the urban area. Only if this can be demonstrated will planning permission be granted for a greenfield site or peripheral development. Local authorities could then place restrictions on the planning permission, so that the developer had to ensure that a certain proportion of all trips were made by modes other than the car. The means by which this would be achieved would be left to the developer, but failure to reach the target levels would result in fines. A development levy could be charged for development on greenfield sites and the proceeds used to help in financing development on brownfield sites in urban areas. At present it is far cheaper to build new houses that are VAT-exempt than to refurbish or convert the existing stock. The VAT rate for conversions was recently (November 2000) reduced from 17.5 per cent to 5 per cent,[5] but, contrary to the recommendations of the Urban Task Force, this revenue will not be reinvested in urban regeneration.[6]

In terms of planning, the principles of sustainable development are clear, namely that higher densities of housing are required and that most new development should be placed in settlements of a sufficient size to support a full range of services and facilities. Where possible, development should consist of mixed land uses and be located near public transport interchanges. In transport terms, such development would reduce the growth in car-based travel, energy consumption and emissions levels. In addition to the physical aspects of planning, it is important to maintain strong communities through increasing local economic diversity, encouraging self-reliance, and promoting social justice, for example, by providing for the housing and living needs of all. The attractiveness of both urban and rural locations should be enhanced through investment in good public services, high-quality urban design and open spaces, and a secure environment. This involves the public sector working in partnership with the private sector and voluntary organisations.

In the recent past, levels of service provision have been very variable, particularly in rural areas. A report by CAG Consultants in 2001 showed that the number of village halls or community centres seems to be increasing, and the decline in schools and public transport services has been curtailed.[7] The other four services identified in the report (public house, permanent shop, post office and general practitioner), however, are still experiencing a continuous process of closure. There is the potential to help solve the problem of isolation or poor access through a mixture of innovative alternatives: the rapid development of information and communications technologies facilitates access to many facilities, co-operative arrangements within rural communities could allow for purchasing larger quantities of goods and services at a cheaper price, and the local shop could be used as a distribution point for goods and services ordered over the Internet.

There is now a general realisation that thinking has to change if sustainable development is to become central to planning, with an accompanying agreement about linking land-use planning and transport solutions, and a creative consideration of alternatives. The decline in some rural services has been reduced[8] and the possibility of a clustering of services under one roof or in a community centre is becoming a reality. The sense of community is becoming more evident as newcomers take on community activities, often supported by planners through a more relaxed application of controls and even through subsidy. All these changes facilitate the move towards sustainable development and the involvement of people in determining the future of their local neighbourhoods.

New horizons and visions

The new millennium has marked a watershed in thinking from central government, with the production of two white papers, one on towns and cities and the other on the countryside,[9] in response to the challenging vision set out by the Urban Task Force in its report 'Towards an Urban Renaissance' (1999), with its goal of creating the 'Sustainable City'. The report establishes the importance of developing the 'Sustainable City' by creating compact urban development based on a commitment to excellence in urban design, and by creating integrated urban transport systems that prioritise the needs of pedestrians, cyclists and public transport passengers. It also recommends that resources be targeted at the regeneration of areas of economic and social decline and at investing in skills and innovative capacity (high technology and information-based activities). It emphasises the need to develop brownfield land and to recycle existing buildings, so that these options become more attractive than building on greenfield sites. Finally, it calls for the use of sufficient public investment and fiscal measures to lever more private investment into urban regeneration projects. The urban renaissance, it is concluded, will not be achieved unless these goals remain a political priority for the next twenty-five years.

Housing in urban areas

Urban areas[10] are seen to be central to sustainable development as they are where some 80 per cent of the population lives and where most new development must be accommodated. The government has set targets for 60 per cent of all new housing to be located within existing urban areas, mainly on brownfield land, and to be developed at higher densities than is currently the case. This imperative is highlighted by the estimated requirement for 3.8 million new dwellings in England by 2021, mainly to accommodate the huge growth in single-person households. Conversion and new construction in the cities will help to reduce outward migration of people, but it is crucial that the new small units are affordable.

There is a serious threat here, as many people in the public sector cannot afford to live in the South East or in London, yet these people are essential to the economy and the diversity of life in cities. If the problem of affordable housing is not addressed, then many health, education, social and public services may break down. Business and industry also depend on lower-paid employees to service their activities. In the past, the public housing sector has accommodated these workers, but increasingly they have become excluded from the housing market and have moved out to the suburbs, where cheaper housing is matched by longer and more expensive commuting. Since the housing market has failed, positive investment is essential to address the lack of affordable, small units in both urban and rural areas. Yet, both the white papers have seriously underestimated the scale and severity of this problem.

The housing problem also reflects regional inequalities and the juxtaposition of wealth and poverty within urban and rural areas. The Regional Development Agencies (RDAs) are primarily concerned with inward investment and the creation of jobs, as this is seen as the means to improve competitiveness. Such strategies do not, however, address issues relating to skills, learning and education, nor to social inclusion, nor to enhancement of the environment. The RDAs need to broaden their remit to encompass these issues and relate them to local needs. This means that planning has to be reformed so that it can encourage or discourage development according to its overall resource intensity, including its impacts on the environment and on the social diversity of local communities. Investment and taxation should be related to a wider concept of employment covering the range and quality of jobs as well as arrangements such as job sharing. This in turn may require a review of the city within an RDA's region, perhaps with the city defined according to journey-to-work areas so that the tax base reflects this wider geography. If local taxation were to relate to the job catchment area of cities rather than the usual geographical boundaries, then those who currently travel from the outskirts and gain from services provided by the city yet contribute very little in local taxes would be helping to finance development in 'their' city. As outskirts are often the most affluent areas, this would bring considerable benefit to the overall wealth of the city. Such a strategy would help to meet the Countryside Agency's goal of one nation linking town and country.

Transport strategy

Despite good intentions, transport and planning policy has not fundamentally changed to embrace the framework for sustainable development. The 10 Year Transport Plan[11] has in fact turned the clock back by refocusing on large-scale projects and road building. The objectives of providing accessibility and reducing travel distances have been forgotten as the dominance of the car driver is reimposed, with mobility, choice and the reduction in congestion re-established as the key objectives. Yet even here the 10 Year Transport Plan does not deliver, as congestion is seen as relating primarily to travel speed rather than to travel time. People do not measure their journeys in terms of the speed they are travelling – it is the uncertainty in the length of time it will take to get to their destination that is the main problem facing all travellers.[12] It may be optimistic to expect radical change over such a short period of time, but unless new thinking and action are forthcoming, then the quality of our urban and rural environments will continue to deteriorate. Perhaps it is through direct action that change will take place, as people become more interested in the issues and more frustrated with the lack of effective measures to deal with them.

Recent developments, from transport disasters and fuel-price protests to adverse weather conditions, have tested the resilience of the transport system to its limits. In the past, the argument has been about increasing speed to save time, and most benefits from transport investments have been intended to save travel time. Yet this rationale has simply led to increasing journey lengths and more travel. Perhaps the sustainable development agenda would argue for a slowing down of travel speeds to encourage shorter journeys and the use of local facilities. This would include the closure of some roads at certain times so that they could be returned to people on foot or bicycle, as has happened in some urban areas, with the Reclaim the Streets campaign. Such measures are equally applicable in the countryside, where roads could be allocated to schoolchildren for cycling or to ramblers for walking. At other times, these same roads could be opened to more general traffic, but with much lower speed limits.

Concepts such as reasonable travel time must become part of the policy agenda. It increasingly seems that the transport network is only as good as its weakest link. At present, high-speed networks that encourage longer-distance travel are more vulnerable to failure, particularly if the organisational structure is fragmented, as is the case with the railways, and if many services are outsourced, as in the freight sector. These weaknesses may be compounded by too great an emphasis on commercial criteria as the main measure of efficiency. Broader social criteria need to be considered when determining who gains and loses from particular decisions and when establishing the consequences for environmental quality.

Sustainable development must be seen as a distinctive and necessary policy agenda, rather than an add-on to the economic agenda. At present, the two white papers are embedded in the Treasury-driven agenda of economic growth narrowly defined, and this will not lead to sustainable development. The debate must be enlarged to include the availability of affordable housing, local facilities and high-quality local transport networks, thus reducing dependence on resources and promoting a high quality of life.

Quality-of-life indicators

One possible means of overcoming the lack of integration between various government policies and thus achieving the goal of joined-up government is the recent development of a series of national indicators of sustainable development, including fifteen key indicators designed to give a broad view on whether a better quality of life is being achieved.[13] These 'Headline Indicators of the UK Sustainable Development Strategy' fall into one of three categories:

- economic: maintaining high and stable levels of economic growth and employment;
- social: achieving social progress that recognises the needs of everyone;
- environmental: achieving effective protection of the environment.

Progress is monitored on all indicators and action taken where the trend is deemed unacceptable. As it takes time for meaningful patterns to emerge, to date most of the changes in the fifteen indicators have been unidentifiable. However, as I have shown, we know that the trends are still continuing, and that they are continuing mainly in the 'wrong' direction. The indicators themselves cannot help change direction, merely reinforce the message that we are not moving towards sustainable development.

Ironically, it is only with the economic indicators that steady positive progress is being made, yet these indicators provide no information about sustainable development, as they are standard measures of economic growth. Some progress is also identifiable in the social indicators. Least progress is being made on the environmental indicators, particularly for the longer term, where tougher targets are needed to address the global issues of sustainable development. For example, the recent Intergovernmental Panel on Climate Change report[14] has proposed a 60 per cent reduction in global greenhouse gas emissions as its best estimate of the scale of change required by 2050 to limit the adverse impacts on the global climate. This means that the developed countries should aim at a target of about a 90 per cent reduction. But the UK government has not moved further than its 20 per cent target reduction for CO_2 emissions! Its most recent set of proposals for more roads, together with the relaxation of the fuel-duty escalator and without any accompanying traffic-reduction targets, all suggest that even this 20 per cent target is going to be impossible to achieve, at least in the transport sector. There are no clear national targets in planning guidance, and local authorities are naturally reluctant to impose local targets individually. A report from CAG Consultants concludes that 'taken as a group, the indicators provide no evidence that the UK has moved significantly closer to reconciling and integrating quality of life with living within environmental carrying capacities'.[15]

Indicators can help in identifying where quality of life is improving or getting worse, but they do not say anything about why a change has taken place, as they cannot be linked to causal factors. The risk is that too much effort is expended in monitoring trends that we already know about, rather than actually tackling the problems at source.

What is more, quality of life is a much more fundamental concept than that suggested by the indicators. At the individual level, it relates to satisfaction, self-fulfilment, security, health and independence. At the community level, it relates to the quality of the built environment, the neighbourhood, social diversity and the local natural environment. At the national level, it relates to economic prosperity, political stability, environmental quality and a fair society. It is impossible to combine all these elements into a series of simple quality-of-life indicators.

Underlying the rhetoric is the suspicion that sustainable development is not the focus of policy, but a 'side show'. The economy itself is the most important concern and takes precedence over all other interests. The increasing reliance on the market, albeit modified by social concerns, is paramount. In second place are the concerns over social welfare and an inclusive society, with environmental issues lagging behind as a very poor third. Unless the balance is redressed, sustainable development is an impossibility, and the indicators used will merely reflect the pre-eminence of economic and some social measures.

Key elements

This essay does not have all the answers, but it asks the essential questions as to whether a sustainable development agenda is emerging from the traditional concerns of planning and transport. There are promising signs, but it is taking a considerable time for the real impacts to become evident.

To address the key elements of the sustainable development agenda effectively from the planning and transport perspective, planners must develop an understanding in each of the following areas:

1   Towns and cities are the key to sustainable development, as it is only here that services and facilities can be provided with the maximum levels of accessibility. The quality of life must be attractive, which means that higher densities of housing need to be balanced with better design, more open spaces and recreational facilities. Perhaps cities will develop as looser agglomerations with multi-centre hierarchies, as envisaged by the Urban Task Force.

2   Sustainable transport can be provided in urban areas through the creation of car-free areas served by high-quality, pollution-free public transport. The best available solutions can be provided with imaginative packaging of transport alternatives,

matched by the use of innovative technology in transport, and the exploitation of electronic technology to maximise home-based work options as viable alternatives to travelling to work.[16]

3 Affordable housing should be an essential part of all new residential development, as it provides both opportunities for individuals to gain access to the housing market and social balance. Single-person households will provide a major new element in society and the majority of them will need to be accommodated in cities.

4 The distinction between town and country is becoming less clear as migration takes place and as traditional rural activities are enjoyed by a wider cross-section of society. Rural areas must change and accept this diversity, providing employment and investment. New city regions based on job catchment areas may help break down some of the barriers and help the move towards a one-nation society.

5 The purpose of green belt policy needs to be reviewed in the context of sustainable development. There is a case for development to take place at the edge of existing urban areas, so that levels of accessibility can be maintained and existing facilities continue to be supported. If development 'jumps' across the green belt, however, then journey lengths are increased and by implication car dependence. Green belts should also be seen as potential recreational land and open access to them should be encouraged.

6 Society is moving away from one based on work to one based on leisure. Leisure now accounts for 50 per cent of all activities and although the home is becoming a high-technology leisure environment, many leisure activities still take place outside the

home through sports, visits and holidays in the UK
and overseas. This brings huge pressures on leisure
facilities, particularly in certain parts of the
countryside. About 1.3 billion day visitors spend
around £8 billion a year in the English
countryside.[17] These pressures will increase and the
challenge will be both to diversify leisure activities
and to minimise the impact of people on the
quality of those activities.

7   The traditional view that travel is only undertaken
because of the benefits derived at the destination
no longer generally applies.[18] Much leisure travel is
undertaken for its own sake and the activity of
travelling itself is valued. People have adopted
lifestyles that are completely car dependent, and
hence are becoming more resistant to price
increases. We must accept that a major rise in the
costs of motoring would not result in the necessary
transfers to public transport, and would bring
hardship to marginal car users and increase the
social polarisation between those with a car and
those without. The only way to address this
problem is through combining planning and
transport strategies so as to reduce the need to
travel.[19] This means that people should live in close
proximity to services and facilities, and demonstrate
a commitment to using them.

8   New technology offers opportunities for distance
learning, teleworking from rural areas, Internet
trading and shopping, and alternative business and
commercial practices. Fundamental changes will
take place in many routine activities, but there will
still be a high value placed on face-to-face contact.
Agglomeration economies will continue to
promote economic efficiency. Some dispersal of
activities will take place, but proximity will still be

an important factor for sustainable development.

9   Planners must change from being reactive to being
    proactive, working with all stakeholders to achieve
    sustainable development. They will need new skills
    and high levels of training. Many developments can
    be left to the market, but planning has an
    important role to play in balancing market forces
    and economic development with the sometimes
    conflicting environmental and social objectives.
    Increasingly, the sustainable development agenda
    should embrace all three dimensions – economic,
    environmental and social – in mutually reinforcing
    strategies. This is the elusive win–win–win concept.

This essay is essentially optimistic in its recognition that sustainable development has the potential to solve many seemingly impossible problems. In a remarkably short time, it has pushed many issues into the centre and its goals have now been accepted – traffic reduction, car-free zones, commuter travel plans, quality design, affordable housing, the reuse of urban land for housing, locally based facilities, high-density living, mixed-use developments, taxation on consumption rather than production, and community involvement. As the Urban Task Force stated, 'we need a vision that will drive the urban renaissance. We believe that cities should be well designed, be more compact and connected, and support a range of diverse uses – allowing people to live, work and enjoy themselves at close quarters – within a sustainable urban environment which is well integrated with public transport and adaptable to change.' The challenge of sustainable development is an exciting one, but we are at present only just realising the scale of the daunting task ahead.

Part Two

Landscapes

Urban landscapes

Simon Jenkins

The lady sitting next to me turned and asked where I lived. I replied that I lived in London, a city to which I felt deeply attached. 'How ghastly for you,' she said. 'How can you live in such an awful place?' On being told that she lived in Leicestershire, I decided to reply in kind. 'How ghastly: how could you live in such an awful place?' She was completely astonished. 'What a rude question,' she said.

The stereotype of country good city bad runs deep in British culture. When in the 1920s the writer H. V. Morton went 'in search of England' he did so on the basis that England meant the countryside. There he found timeless values, the essence of nationhood and the racial rootstock of Empire. On every side, he saw such values threatened by town, by urban machines, urban crowds, urban immigrants and radical urban values. Writers from William Cobbett and George Eliot to John Betjeman and Raymond Williams have found in the town–country divide a metaphor for every sort of social conflict and cultural upheaval. Nothing in the debate over the fate of the countryside suggests that this dichotomy has diminished.

In the late 1990s I went on a series of journeys to discover the finest parish churches of England (time prevented a wider survey of all Britain). I returned from each visit elated by the churches but gloomy at the state of the English landscape as a whole, rural and urban, a landscape I came to regard as a seamless web of conflicting social and economic pressures. The country was the most vulnerable, though by no means the most abused. The country I regard as the nation's most valuable yet intangible long-term asset, one that economists tell us is near impossible to 'monetise' and therefore protect.

We are told by the Council for the Protection of Rural England (CPRE) that between 1980 and 2000 more open land vanished under development as a percentage of the whole than at any time in the twentieth century. It vanished because town and country planning virtually collapsed under pressure from commercial rather than state development, accelerated by pro-development changes in planning procedure under the Thatcher government. Time and again I found local people apologising for some outrage inflicted on their village or town – a bypass, a hypermarket, a row of farm silos, a commercial estate, acres of executive homes – as if theirs was a unique misfortune. They were genuinely shocked to find that their experience was universal. But then few Britons have wide experience of their own country. They travel along narrow corridors and travel extensively only on the Continent.

Circumnavigate any large settlement in England short of a metropolis and you will see round it a spreading stain of un-coordinated, unplanned suburban development. Round Exeter or Cirencester, Cheltenham or Ely, Nottingham or York, a fringe 'green' activity such as playing field, pony club or farm depot wins planning permission to move farther out and, on the grounds that it is already semi-developed, is allowed to sell on to a speculator with planning permission. Wherever there is 'soft' land near a garage, hypermarket or motorway junction, there is pressure to infill round it. A hard-pressed farmer with friends on the council wins a caravan park or industrial estate. The process is now utterly remorseless. Nor does it have anything to do with relieving urban housing stress. I have yet to see a new estate in Bedfordshire or Buckinghamshire, Cambridge or Somerset that is anything other than for car-borne commuters. England may have the lowest amount of green space per head of any big country in

Europe, but it continues to lavish what is left on the lowest housing densities.

England has in the last quarter of the twentieth century gone down the same route as much of the northeastern United States and 'suburbanised' its countryside. The first stage is the move from city to country of the mobile rich, with a consequent rapid rise in house and land values. This prices local people out of settled communities and imposes a burden on local taxpayers for new infrastructure. Newcomers demand changes in building-use classes – farming to warehousing, shopping to housing, garage to supermarket. Above all, the new population is car borne, causing the collapse of local retail markets and other services because needs are now satisfied over far greater distances. The countryside becomes a giant dormitory, in the south-east much of it occupied for weekends only.

To all this the fate of farming is irrelevant. The stimulus to rural change comes not from varying agricultural incomes, though that can undoubtedly push a small farm into crisis, but from soaring land values and loosening planning control. Indeed the fastest rate of urbanisation – the CPRE's 'another Bristol every five years' – was during the 1980s, when agriculture was enjoying a prolonged burst of prosperity. When planning permission on a meadow in Essex can add ten or twenty times its value, the state of the rural business cycle is a marginal consideration. Farming is becoming a residual activity in the space in between urban areas. It is a landscape in mothballs, a capital asset awaiting exploitation. There is no shortage of exploiters eager for the opportunity.

The countryside has a feature that the urban landscape lacks. In Mark Twain's words, they don't make it any more. While countryside can be turned overnight to town, town cannot be turned overnight to countryside. This simple market asymmetry underlies the town–country debate and should inform its politics. Countryside offers a perfect instance of what economists call an external benefit and one that is near-impossible to quantify. Destroying it profits both seller and buyer. The loser is the community as a whole. The benefit of the English countryside cannot be quantified and charged. It is revealed only in opinion polls that always put it high in the 'most valued elements of the British way of life'. Yet only a few locals notice as each specific acre goes under concrete. They merely pack their cars and head south to the Continent – to France, Italy and Spain, where geography has left plenty of open space still to enjoy. Britons now head overseas for their leisure much as Americans head west, in the simple quest of sheer space.

The prospect of England outside its national parks as essentially a large suburb may seem far-fetched to anyone flying low over the country today. But if the development is in that direction then it will eventually happen unless specific steps are taken to halt it. The future exists. It can be seen by flying over New Jersey or Connecticut, Phoenix or the 'metroplex' of Dallas–Fort Worth. It can be seen by flying round most of the coastline of southern England or over the fastest-growing metropolis in England, that of Bristol. The coastline is now heavily developed, with caravan and bungalow development up to the fringes of every scrap of publicly owned land. The Kent, Sussex and Hampshire shoreline is vulnerable to ribbon development everywhere that it is not specifically protected. It is a template for the future of the countryside.

The pace and nature of building in the Home Counties over the past quarter century are similar to those in New England, but with a difference. In New England local communities are permitted to stipulate plot ratios. They can keep their dormitories to one-acre, four-acre, even ten-acre plots. They can opt for rurality, thus reducing some of the strain placed on the road network and other services. In Britain central government insists on housing targets being met by county authorities, even enforcing them in the courts. This naturally encourages high densities, though not high enough to vitiate the need for a car for almost all journeys. Though this pressure is less in the north of England, any drive from Merseyside through south Lancashire over the Pennines to Leeds offers a spectacle of the most

1: The Old Royal Naval College at Greenwich opens up views from the Thames to the park and to Blackheath beyond. (Introduction)

2: Award-winning flats at Battersea overshadow St Mary's Church and present a cliff-face to the Thames. (Introduction)

3: Artist's reconstruction of an interglacial scene of approximately 120,000 years ago on the south Wales coast, showing five mammal species: hippopotamus, straight-tusked elephant, rhinoceros, red deer and spotted hyena. The trees include maple and oak.
(The effects of climate change)

4: The devastated Holbeck Hall Hotel after a dramatic coastal collapse near Scarborough on the Yorkshire coast in June 1993.
(The effects of climate change)

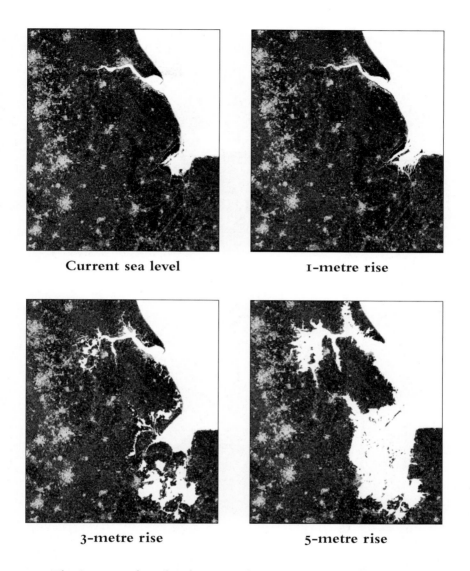

Current sea level

1-metre rise

3-metre rise

5-metre rise

5: The impact of sea-level rise on the eastern coast of England. A digital elevation model has been used to create a simplified and approximate visual representation of the results of rises of 1, 3 and 5 metres above current sea level. (The effects of climate change)

6: 'Over the period 1989–99, motorway traffic has grown by 42 per cent in Britain.' Traffic at a standstill at junction 24 on the M1 motorway, 2 November 1999. (Sustainable development)

7: The spreading stain of suburban development around Ely, Cambridgeshire. (Urban landscapes)

8: Countryside or suburb? Permanent coastal caravan park at Bigbury-on-Sea, Devon. (Urban landscapes)

9: The interface hosts much informal recreation, as here on the northern edge of York. (Edgelands)

10: This wildlife-rich gravel pit at Molesey Heath in south-west London has since been partly filled in. (Edgelands)

11: The interface is rich in industrial archaeology, as these derelict buildings at Llanelli show. (Edgelands)

12: Industrial regeneration, as here on the Isle of Thanet, has moved from the inner city to its edge. (Edgelands)

13: Some more pastoral scenes remain, relics of an earlier landscape interspersed among modern intensive cultivation. (Lowland landscapes)

14: Bulk agricultural production takes place under factory conditions; all cover has been stripped away to minimise costs of production. (Lowland landscapes)

miserable and extravagant sprawl. What Nikolaus Pevsner described in the 1960s as 'the most densely populated landscape in Europe', that of south Lancashire, has probably tripled its occupation of land since then.

The impact of this pace of change on the existing townscape is all too familiar. The decay of inner cities, not just in the commercially depressed north, has been stark. While government ministers talk of Hampshire's 'need' for 40,000 new houses over the next four years, a similar number were reportedly lying empty in Portsmouth and Southampton alone, quite apart from those now severely underoccupied because security of tenure means the elderly are holding on to more space than they need. British urban housing is a case of total market failure. Government confuses need with demand. It predicts household formation, divorce rates, the 'need' for second homes or out-of-town homes, and then requires that the planning system responds accordingly. There follows a fatuous debate about the 'appropriate' split between redeveloping green, brown or plain black land.

The result has been the greatest stupidity in the history of public planning. For almost half a century, public money has been poured into demolishing and rebuilding the centres and inner suburbs of English cities. Similar sums have then gone into subsidising agriculture. Then, as commercial pressure decentralises demand from unattractive cities to attractive countryside, huge sums have gone on supplying the roads, schools, power and water supplies required by that development. Then more money has been spent rescuing the resulting decay of the cities, seeking to redirect development back into them. It is as if the taxpayer was running back and forth across the boundary between urban and rural England scattering money. The outcome has been catastrophic to both rural and urban amenity. Any more robust profession than planning would have held impeachment hearings.

There is no 'planning need' to develop any of England's remaining green land. There may be market pressure to do so. There may be eager buyers of property and eager sellers of meadows. But there is nothing that could be termed a need. The very concept is drawn from planning's early socialist past: 'to each according to his needs'. Planners may feel under a political or commercial obligation to respond to rising prices by supplying a scarce commodity that is under their custodianship. But that is nothing to do with need. I could as well refer planners to the need for urban renewal, or the need to fill existing empty property in city centres. And I could certainly cite an overwhelming need that can only be met by the specific skill of planning and the power vested therein, which is the need to protect what remains of England's open spaces. Other than through the agency of public bodies such as National Parks and the National Trust, that need can be met only through regulation. And that regulation, at present, is in a state of demoralised chaos.

If the impact of planning failure on the rural landscape has been easy to see, that on townscape has, I believe, been perhaps less visible but no less painful. The fate of most of England's town and city centres over the past half-century has been tragic. The failures of health, education and crime policies are, at least, short lived and correctable. Those of planning live through generations. Fault has lain in two quarters, the first political, the second aesthetic. Those charged with the reconditioning of Britain's towns and cities after the war were both left wing and self-confident. They held that capitalism had failed the city and that landscape should be refashioned for the benefit of the urban working class. The middle class might seek solace in the suburb. The city was the New Jerusalem of the urban proletariat. I watched the bulldozers moving across Walworth in South London and across Moss Side in Manchester in the early 1970s. The destruction was greater than anything wreaked by war. It was as if a class memory was to be eradicated. Such an attitude, well described in Andrew O'Hagan's recent novel, *Our Fathers*, was universal throughout urban England.

This political thrust was given added impetus by the aesthetic of the Modern Movement in architecture, also political in much of its motivation. It encouraged the new urban leadership to identify old buildings with Victorian and Edwardian capitalism, and socialism with technological modernity. I sometimes wonder if my response to the great post-war clearances is exaggerated, in part through having witnessed them and having seen so much of the consequence. But the change wrought to urban Britain roughly from 1955 to 1980 was devastating. It has been well documented by Oliver Marriott (*The Property Boom*) and Lionel Esher (*The Rebuilding of Britain*), as well as by the many 'Then and Now' picture books that adorn local bookshops. The only parallel with the destruction of urban Britain after the last war (other than the destruction of rural Britain) was that which occurred in the Marxist cities of Eastern Europe.

Thus it was estimated that a third of the timbered houses of Coventry were destroyed in the Blitz. The city fathers subsequently destroyed almost all the rest. The unique row houses of Great Yarmouth were half destroyed by bombs. Almost none survive today. Bombing was used all over England as an excuse for the eradication as 'slums' of houses that, in West London, were being converted into smart residences for the new rich. Bulldozers wiped out areas of cities that, today, would form the historic districts rightly regarded as magnets of urban renewal, for tourists and local citizens alike. Warsaw, Budapest, Hamburg and Tours restored, even rebuilt, their historic cores. Plymouth, Southampton, Portsmouth, Bristol, Coventry wiped theirs out. They seemed almost to welcome the bombs as presaging the doom of capitalism and the chance to begin again.

Almost all the post-war 'fresh starts' have been failures, presenting architecture with what should be the most glaring professional malpractice suit of the twentieth century. While the medieval and eighteenth-century shopping districts of York, Oxford, Bath and London's West End have been able to adjust and update themselves to changing market circumstances, such prize-winning post-war set pieces as Bristol's Broadmead, Plymouth's Armada Way, Birmingham's Bullring, Liverpool's St George's and Manchester's Arndale have failed and many have had to be demolished. The same has been true of the associated high-rise housing that replaced cottages and terraced estates. System-built housing in Sheffield and Leeds has been demolished, as have the high-rises of Liverpool and Glasgow. They had become uninhabitable hells.

My father's birthplace of Dowlais in Glamorgan was the cradle of the south Wales industrial revolution. Home to the mighty Dowlais Ironworks, it was an industrial monument as important as, and far more dramatic than, Coalbrookdale in Shropshire. It would today be a World Heritage Site. In the 1970s every structure was demolished, every furnace, every kiln, every mill, every workshop and office, the high street, the shops, the schools, the terraced housing, everything went. An orgy of destruction descended on Dowlais as if a Stalinist had sought to wipe all memory of it from the face of Wales. Nothing that recalled the old was allowed to remain. The housing estates that have come instead are characterless and bleak. Vast sums were needlessly spent on what would today have been a tourism and employment magnet, to create a wilderness. These crimes were committed in the past quarter century.

The architects and social engineers of the 'new towns' made extravagant claims about what these brave new urban areas could achieve. Yet Corby and Skelmersdale, Cumbernauld and Washington became bywords not for high-tech employment and social cohesion but for 'new town blues' and crime. It was astonishing that officials calling themselves planners could have driven not thousands but millions of people out of their familiar and restorable homes in central Liverpool and Manchester, to spread over the green belt and countryside of south Lancashire in the 1970s and 1980s. I watched people in tears being bussed like Balkan refugees away from friendship and family ties and from any hope of informal employment. The city centres were denuded of economic activity and the countryside destroyed into the bargain. I repeat, I regard the people who ordered these clearances as guilty of crimes.

These post-war catastrophes of political and aesthetic

judgement have vastly complicated the revival of the British economy in the last quarter of the twentieth century. They continue to scar generations of young people whose community ties have been shredded and who have turned to crime and drugs as a result. The underclass estates of Britain remain the most sociologically intractable in Europe. Nobody who has not travelled the estates of inner Liverpool, Manchester or Leeds should question these conclusions.

Now we are charged with trying to rectify these great errors, to find optimism among a wreckage far worse than the wreckage of war. What is most instructive is to discern what already forms the basis of recovery. It is invariably such relics of an older urbanisation that survived the destruction. The liveliest urban neighbourhoods today – and thus usually the ones with the fastest-rising land values and the most magnetic retail zones – are all old. They are the Gas Basin and jewellery quarters of Birmingham, the Calls and canal district in Leeds, the historic waterfront in Newcastle, Albert Dock in Liverpool, the Waterfront in Bristol, the New Town and Leith in Edinburgh. In London they are Mayfair, Kensington, Covent Garden, Notting Hill, Camden Town and Shoreditch.

There is now a clear relationship between property value and conserved townscape. In 2001 commercial rents in the City of Westminster finally overtook those in the City of London. The reason is beyond doubt. Westminster is almost entirely a conservation area and is a pleasant place in which to work. The City has abused its conservation areas and disregarded its street environment, only pedestrianising any of its streets under pressure from IRA bombing. It is increasingly not a pleasant place in which to work. The market is talking. And the market demands that the 'externality' of an attractive urban environment be protected to maintain the internal value of developable land. That requires planning intervention. This applies as much to the overheated property market of inner London as it does to the desperately impoverished markets of the urban north of England. The beauty of a conserved townscape is as relevant to downtown Bradford and

Stockport as it is to Camden and Kensington, Hoxton and Bermondsey.

The case for rescuing Britain's urban landscape is thus part and parcel of the conservation of what remains of the countryside. Government has yet to recognise this. In 2001 the government published two white papers, one for the future of towns and one for the future of the country, thus perpetuating a harmful historic dichotomy. Both are part of the same planning challenge, to seek a landscape in which land values are harnessed to the usefulness and wider enjoyment of all. Only by making towns more attractive, by increasing urban densities and protecting historic neighbourhoods, can pressure on the existing countryside be abated if not altogether relieved. All land is scarce.

Planning is the oldest of political activities. The biblical Book of Ezra is full of rules for the rebuilding of Jerusalem. Tudor monarchs were obsessed with the coherent planning of London. So too were the Restoration authorities, the authors of the Georgian and Victorian building acts and the bold innovators of town and country planning between the wars. More recently planning has come to embrace not just statutory protection but fiscal incentives and subsidies designed to point development in certain directions. These tools have not always been productive, and often counter-productive, to good planning. Yet it is inconceivable that the challenge facing the English landscape can be met without them.

Policy for the countryside is for other essays in this book to consider. As far as town planning is concerned, outside existing conservation areas, Britain is still in the Dark Ages. The experience of walking round French, German or Scandinavian towns and cities, round Barcelona or Perugia, Amsterdam or Geneva, is sobering for any Briton. How do they get it right and we get it so wrong? They are ahead both in the use of fiscal and planning rules guarding their skylines and building materials, and in the design and management of public spaces, pavements and squares.

One reason is that the structure of local government on the

Continent favours civic pride. It encourages an actively involved citizenry, willing and allowed to tax themselves to better their environment. In rate-restricted Britain such taxes are not permitted by central government. Most continental cities also recognise that active intervention is required to prevent business fleeing the city, taking people and money with it. This means re-establishing town centres as places of public resort. It means aggressively removing the excrescences of the post-war period where these damage the appearance of the city and impede the regeneration of its streets. It means resisting the craze for macho high buildings, a sort of banana republic obsession of city planners who feel obliged to show their virility.

A revived civic dignity and leadership are an essential pre-condition for a revival of confidence in urban England. More than that, there needs to be a specific programme of townscape improvement. Nothing would do more for the appearance of the once-fine city of Newcastle than the simple removal of the six high-rise blocks inserted in the 1970s and 1980s. The same goes for the removal of the multi-storey car parks that blight every vista in downtown Bristol. The urban motorways built close to the centres of Glasgow, Birmingham and Nottingham are probably irreparable. But unneeded gyratory traffic systems continue to blight the centres of Leeds, Southampton, Worcester, Bedford, Leicester and countless county towns. These can be rectified, but only with a will.

The biggest challenge of all is simply one of design. Given the huge sums available for urban renewal, public and private, it is astonishing how few successful works of modern town planning have been built that are likely to last. The only ones I can cite, such as that of central Birmingham, depend heavily on the rescue and renovation of existing 'obsolete' buildings. The new centres of Aylesbury, Basingstoke or Hemel Hempstead, centres of rapid commercial growth, are crude monstrosities, already racked with crime. Scale is the enemy of intimacy, yet scale is easy to finance and simple to build. So bigness is all. Huge blocks of office and retail use now form alien citadels in Sheffield,

Chesterfield and Colchester. The approaches to Ipswich, Portsmouth and Wolverhampton are dominated by commercial blocks of extraordinary ugliness, many of them created in the past two decades. The architecture of arrogance is ubiquitous, the architecture of deference rare.

I am pessimistic, but in this matter implacable. The damage inflicted on English towns over the course of my lifetime in the name of prosperity and social engineering has impoverished them beyond belief. I resent having to travel abroad to see clean, well-ordered, prosperous and popular towns and cities, cities in whose central areas people of all classes will want to work, live and enjoy themselves in the evening. Victorian visitors to London used to comment on how genteel were the city streets, how the English spent their time in neat private boxes, at home with their families. This introversion was considered a human virtue, if not a civic one.

Modern English townscape is the embodiment of the celebrated contrast between private affluence and public squalor that was drawn by the American economist, Kenneth Galbraith, in describing America's public realm. The city boxes and their genteel streets have been demolished and their inhabitants moved out of town. English introversion has been translated from city to suburb and on out into the country. Rescuing England from this process, rescuing its urban and rural landscapes and restoring to each its appropriate function, is the greatest challenge facing politics today.

Edgelands

Marion Shoard

Britain's towns and cities do not usually sit cheek by jowl with its countryside, as we often casually assume. Between urban and rural stands a kind of landscape quite different from either. Often vast in area, though hardly noticed, it is characterised by rubbish tips and warehouses, superstores and derelict industrial plant, office parks and gypsy encampments, golf courses, allotments and fragmented, frequently scruffy, farmland. All these heterogeneous elements are arranged in an unruly and often apparently chaotic fashion against a background of unkempt wasteland frequently swathed in riotous growths of colourful plants, both native and exotic. This peculiar landscape is only the latest version of an interfacial rim that has always separated settlements from the countryside to a greater or lesser extent. In our own age, however, this zone has expanded vastly in area, complexity and singularity. Huge numbers of people now spend much of their time living, working or moving within or through it. Yet for most of us, most of the time, this mysterious no man's land passes unnoticed: in our imaginations, as opposed to our actual lives, it barely exists.

When we think of the land of Britain we think of town and village, countryside and coast. Our image of Kent is still one of towns, wealden or coastal, neatly demarcated from downs, orchards and fields. When we think of Scotland we think of Edinburgh Castle and heather-clad hills. We are, of course, also well aware of the great conurbations. But not of the edgelands.

The apparently unplanned, certainly uncelebrated and largely incomprehensible territory where town and country meet rarely forms the setting for films, books or television shows. As we flash past its seemingly meaningless contours in train, car or bus we somehow fail to register it on our retinas. When we deliberately visit it, this is often for mundane activities like taking the car to be serviced or household waste to the disposal plant, which we choose to discount as part of our lives. If we actually live or work there, we usually wish we did not. Public authorities tend to have a similarly negative attitude towards these interfacial areas.

Sometimes these areas are so little acknowledged that they have not even been given distinctive names. Over the past twenty years on land north of Bristol an employment and shopping centre has grown up over what was agricultural land.[1] This development has become so large that it now rivals in size the central business district of Bristol itself. The close proximity of three motorway junctions has attracted a science park, the procurement headquarters of the Ministry of Defence (one of the biggest office developments in Europe, with a workforce of 6,500), British Aerospace, Rolls-Royce, the European headquarters of Orange and the European headquarters of Hewlett Packard, together with a shopping mall offering as much retail space as the main shopping centre in the heart of Bristol. The University of the West of England is also to be found here. Yet the whole area, which provides employment for 34,000 people and has an employment capacity of 66,500, is known only as the 'North Fringe'.

In other places, de facto settlements are known by names that fail to reflect their current character. Until the mid-1980s, 'Westwood' in the Isle of Thanet, east Kent, was the name for the crossroads between the main roads serving the area's three long-established coastal resorts – Margate, Broadstairs and Ramsgate. Today, it denotes a vast developed sprawl in the centre of Thanet, featuring massive retail and business parks, a motor-cycle training centre, a riding establishment and a branch of the

University of Kent, all of these rubbing shoulders with cabbage fields, old, isolated farmhouses and farm-workers' cottages, and a closed mental hospital; the road names bespeak the juxtaposition of old and new – 'Poorhole Lane' beside 'Enterprise Road'. Yet Westwood has no centre and, in an area long denuded of tree cover, none of the character that the place name suggests.

Once the suburbs were also a netherworld whose existence went unacknowledged. These sudden accretions to ancient settlements, often unwelcome to opinion-formers, also somehow failed to signify, long after they had become far bigger than the towns that spawned them. Eventually, however, the suburbs won recognition. They acquired first noisy detractors, who at least paid them the compliment of noticing their existence, and then their own enthusiasts and advocates. Some came to be admired while others were deplored. They infiltrated jokes, literature and folk wisdom, and entered the consciousness of us all. They are still not universally loved, but they enjoy a place in our world view and their function is respected. Victorian north Oxford is not yet thought of as quite such a vital part of the nation's fabric as the city's medieval high street, but it has devotees hardly less enthusiastic. It is possible to imagine future generations coming to view the Victorian and Edwardian suburbs of some of our towns and cities as more noble than their often far more ravaged original centres. Apparently monotonous suburbs, like the 1930s additions onto North London, are also seen to be a valid part of our total environment.

Yet so far no similar process of recognition has even begun to be extended to the urban–rural interface. How could it? These jungles of marshalling yards and gasometers, gravel pits, waterworks and car scrapyards seem no more than repositories for functions we prefer not to think about. Their most obvious components are things we have been brought up to think of as blots on the landscape. The apparently random pattern in which they are assembled seems to defy the concepts of orderly planning by humans and of harmony in nature. And yet, there are reasons why the interface could and should follow the suburbs from the dark pit of universal disdain into the sunlit uplands of appreciation, if not acclaim. It, too, has its story, and one that is at least as interesting as that of the suburbs. Perhaps, if we can strip ourselves of our negative preconceptions, we may find it as appealing as well.

There is another reason why the edgelands demand our attention. We may not notice it, but it is here that much of our current environmental change, and in particular the development of large-scale retail, business and industrial premises, is taking place. In the United States there is growing awareness of the development of what has been termed 'edge city' and the effect this is having on the geography and economic and social profile of entire regions.[2] Here in Britain planning still focuses on the problems of towns and the challenge of the countryside. But if we fail to attend to the activity in the interface we forfeit the chance not only to shape that change but also to influence the effects of it on other parts of the environment, such as town centres.

Tasting a landscape

How can we set about the task of evaluating an ignored landscape? In 1965, two scholars at University College London, David Lowenthal and Hugh Prince (one from the United States, the other British), tried to discover the common features of landscapes most admired by those who lead landscape taste in England. They carried out their survey by identifying views expressed in literature, speeches, newspaper articles, letters and at public hearings. Although nobody since Lowenthal and Prince has attempted to cover the same ground, there is no reason to suppose that their conclusions do not remain valid.

English people like to see diverse but calm landscapes in which various elements are arranged in an orderly manner considered picturesque, they reflected. 'What is considered "essentially English" is a calm and peaceful deer park, with slow-moving streams and wide expanses of meadowland dotted with fine trees. The scene should include free-ranging domestic animals ... When it is arable land, hedgerows and small fields are usually obligatory.'[3] They discovered that the English do not like geometrical, obviously pre-planned landscapes, but they do applaud tidiness and neatness. As part of a nostalgic desire to put the clock back, English people like to see evidence of age in their landscapes – old buildings, old trees (native deciduous rather than alien conifer) – and they flock to landscapes rich in historical association. At the same time, they dislike evidence of the present-day and in particular anything obviously functional.

What is immediately apparent about interfacial landscape is that it embodies the antithesis of the characteristics likely to appeal to the appetites identified by Lowenthal and Prince. 'The countryside beloved by the great majority is tamed and inhabited, warm, comfortable, humanized,' they wrote. The edgelands, in contrast, are raw and rough, and rather than seeming people-friendly are often sombre and menacing, flaunting their participation in activities we do not wholly understand. They certainly do not conform to people's idea of the picturesque by presenting a chocolate-box image, suitably composed and textural. On the contrary, they seem desolate, forsaken and unconnected even to their own elements let alone to our preferred version of human life. Tidiness is absent: here no neat manicured lawns with sharply demarcated edges are found. If there is grassland, it is likely to be coarse and shaggy. It may be grazed by perhaps a few sheep enclosed by derelict fencing leading to tumbledown farm buildings. If ungrazed, it will be swamped by a riot of wild, invasive plants that seem to over-run everything in their path: fragments of tarmac, wrecks of cars and derelict buildings. Lowenthal and Prince tell us that the English applaud the absence of litter in a landscape: the interface sucks in the detritus of modern life. Not only is litter and household waste casually dumped there because it is closer at hand than the hedgerow and less effectively policed than the high street, but formal waste-processing operations, such as car crushing, sewage treatment or waste transfer, are often located deliberately here, lest they despoil preferred environments. Edgeland rubbish tips may eventually be grassed over, but the cosmetic treatment of unsavoury artefacts is considered less necessary in the interface than in either town or country. Our gardens, whether in city or village, are expected to be extravagantly manicured. Allotments, the interface's bastard half-child of agriculture and horticulture, put on no such show. Laboured over at least as diligently as gardens, they nonetheless flaunt 'unsightly' infrastructure that would be unthinkable in a garden: old piping used as plant supports, soggy carpets keeping down weeds and tool sheds built imaginatively but untidily out of packing materials.

Local councils commonly require far lower standards of design for new buildings in the interface than they do elsewhere: effectively edgelands have become the lowest grade of landscape in UK landscape conservation terms. The facades applied to urban and rural landscapes to help them satisfy the tastes identified by Lowenthal and Prince are considered unnecessary in a landscape that nobody is actually trying to pretend is other than it is. We do not expect the hypermarkets or the giant factory sheds of the edgelands to blend in with the local vernacular architecture, and where any tree planting is stipulated by councils as landscaping it is usually only the barest minimum. As if this were not enough, the interface also challenges another key characteristic of favoured English landscapes: dislike of the functional. For it embodies naked function more than any other type of environment. Its great warehouses, car scrapyards, rubbish tips, electricity sub-stations, motorway interchanges, welding and car-breaking units pursue their purposes without even a nod to aesthetic conventions.

A useful space

The characteristic appearance of the interfacial landscape is matched by the characteristic forms of land use it reflects. These uses gravitate to the interface for a combination of reasons. Sometimes the cause is obvious: motorcycle training centres, for example, are noisy yet require easy access to centres of population. Drive-to retail units would be inside towns if retailers could find there the floorspace, the parking area and the consequent relaxed planning regime they require. As these retail parks emerge outside towns the road network adjusts to provide access to them, and a chicken-and-egg situation arises whereby interface sites become more attractive to car users and therefore to retail developers. As shopping is coming to be seen more and more as a leisure activity than a chore, superstores are coming to be surrounded by other types of leisure development, such as restaurants and nightclubs. Business parks, distribution depots and housing estates also spring up in the interface, often around the bypasses and motorway interchanges that it provides. Here they sit cheek-by-jowl with the uses we have expelled to the interface because we consider them unsuited to more polite environments − gypsy encampments, rubbish tips, recycling centres, mental hospitals, sewage works and telecommunications masts.

The characteristic activities of the edgelands are also arranged in a distinctive way. Unlike a garden city, say, or a Victorian suburb, interfacial areas are not designed from scratch. They assemble themselves in response to whatever needs are thrust upon them, and in whatever way they can. Motorway interchanges, scrapyards and electricity pylons were not thought of when most settlements were conceived: they were tacked on to the edge of settlements because that was as close as they could be brought, mingling with any gravel excavations, mills and factories that were already there. So in the interface we see history as in the stratified layers of an archaeological site. This characteristic makes the interface intrinsically casual. And unofficial. Even stores appear dumped individually, often surrounded by their own extensive car parks, rather than linked together as in a high street.

Did the settlements of the past spawn a similar penumbra of land servicing the urban area with less glamorous functions?

Often they seem to have done, though little research has been carried out on this subject or on other areas that might give us clues. For instance, we know little of the way in which past societies dealt with rubbish, or even of the nature of the waste that remained unrecycled. York in Roman times is one place that has been studied. There, archaeologists have found evidence of four uses just outside the boundary of the Roman fortress and of the *colonia*, the deliberately planted civilian town across the River Ouse. These were pottery and tile manufacture in kilns (which would have been hot, smelly and noisy), military training at so-called practice camps (which required considerable amounts of space), rubbish deposition (collections of broken pottery and animal bones in fairly large amounts suggesting organised rubbish deposition) and, particularly interesting and special to the period, burial.[4] Since the advent of Christianity to Britain, burial has taken place mainly within settlements, but it was the practice of the Romans to bury their dead outside their settlements.[5] This was probably partly for reasons of public health, and partly because they considered that the dead should have a separate world to inhabit. Elsewhere around Roman York the interface would almost certainly have been devoted to the production of food to supply the garrison and the civilian population.

Characteristically interfacial landscape does not occur today around every settlement. There is plenty of altogether rural land rubbing shoulders with settlements, often quite large ones. It is equally true that land with the distinguishing features I have identified as typically interfacial is not found exclusively on the present-day border between town and country. Although yesterday's interfacial zones are often swallowed up by subsequent building, sometimes they survive as edgeland within built-up areas. Nonetheless, the present rural–urban interface is the expanding landscape of our own age. Disregard of the edgelands means that data about it is largely absent. But here is one statistic: in the green belt around York, a quarter of the supposedly protected green belt within one mile of the edge of the built-up

area was developed between 1966 and 1996, according to calcu-
lations by Michael Hopkinson of the University College of
Ripon and York St John.[6] Every single development reflected in
this statistic would have involved the overturning of the local
councils' green belt presumption against development. Imagine
how much faster development must have been taking place
(though figures are not available) in fringe areas where no such
presumption existed.

Wild at heart

Though usually either unloved or ignored, edgeland does fulfil
vital functions. One of the less obvious is the role it is coming to
play as a refuge for wildlife. The 13.8 hectares of Molesey Heath
are the highest point in an interfacial finger of land on the edge
of southwest London extending from Hersham in the south to
East and West Molesey in the north. This area takes in raised
reservoirs, an industrial estate, an abandoned sewage works, an
equestrian centre, a council housing estate, piecemeal private
housing (some of it on plotlands), and a camping and caravan
site. The Heath is a recent landscape feature: the result of gravel
extraction followed by infilling of rubbish. Stepped paths climb
the 50 feet to the top, and a surfaced track runs round its cir-
cumference, but little else serves the visitor and there are no
elaborate interpretation boards. A handful of horses belonging
to local travellers prevent encroaching fennel, blackberries and
hawthorn from replacing the purple, pink, white and yellow
jungle of common mallow and goat's-beard, tufted vetch and
yellow vetchling, musk thistle and wall rocket – 311 species of
flowering plants and ferns in all. This diversity of plant life is
partly the result of the juxtaposition of many different sorts of
soil, for instance through deposition of builders' rubble, in a way
which would be highly unlikely in Nature. The fact that the
travellers' horses frequently turn up bits of broken glass does not
affect the nature conservation value of the site one iota, nor does

it worry the grasshoppers and crickets that make use of the little open areas, whence they provide an almost deafening aural background in high summer. Indeed, rubbish infilling has raised the land here to such an extent that Molesey Heath offers fine views – as far as Canary Wharf to the north-east and the North Downs to the south. Were this topography the result of some natural process, Molesey Heath would no doubt be celebrated as a viewpoint. And were the lake that lies below it nestling into a sandy cliffside a natural phenomenon rather than the result of groundwater accumulation after gravel extraction, it would not now be in the process of being more than half filled in with rubbish. So disregarded is this lake that it bears no name, yet here, before infilling and thus the partial destruction of the lake started in 2001, the water was of an extraordinary clarity and afforded views of a wealth of submerged green vegetation. Great-crested grebes, kingfishers, coots, moorhens, sedge warblers, frogs, toads, newts and twelve species of dragonfly bred here, while two types of bat would flitter overhead at dusk.

Molesey Heath is not unusual in the rich diversity of its plant and animal communities. Other interfacial wildlife hotspots include the land around Beddington Sewage Works near Croydon, which is one of the top birdwatching sites in southeast England, while all five species of grebe occurring in Britain from the familiar great-crested to the rarely seen red-necked have been recorded at Stonar Lake on the edge of the small town of Sandwich in Kent. This extensive, neglected water, the result of quarrying a hundred years ago, overlooked by a row of rough pines on one side, is not so much as glimpsed by the hundreds of thousands of tourists attracted to this historic town each year. The massive hulks of a pharmaceutical works and a derelict industrial estate strewn with corroding car bodies and litter loom over the other sides; warnings of guard dogs on patrol, which punctuate a crumbling fence, deter the less adventurous from picking their way through the jungle of bushes on the water's edge, which in early spring conceal drifts of silky, yellow coltsfoot.

Clifton Backfields on the northern fringe of York typifies the wildlife value of edgeland space, and represents a rare example of this value actually being appreciated. The Backfields is an unprepossessing stretch of 12 hectares of rough vegetation and scrub close to housing estates, some of them notorious, and partly overlooked by corrugated asbestos hangars and massive, new shed-warehouses. The wanderer here finds him or herself in a varied and beautiful world. There are little patches of woodland with oak, ash, blackthorn, hawthorn, wild rose and apple trees. Deep within these thickets lie occasional stretches of concrete track, their sides covered by carpets of bright-yellow creeping cinquefoil and biting stonecrop. These tracks are a legacy of the last intensive use of the site – by the Army and Air Force during and after the Second World War for, among other things, the Berlin airlift. Most of the Backfields (or the Backies, as it is known locally) consists of fields lightly grazed by rough horses – grassy ridge-and-furrow fields, which bespeak cultivation in medieval times though hardly at all since, and old hay meadows. Here betony, great burnet, red fescue, pepper saxifrage, knapweed, and many other plants characteristic of such habitats flourish. Fifty years ago most farms in this area would have included such flower-rich pasture and hay meadows; now in the countryside they have been almost totally displaced by intensive arable land or stock-rearing monoculture. But in the Backies, away from both plough and agrochemical, the wild flowers and grasses of yesterday's Yorkshire countryside live on.

After the war, ownership of Clifton Backfields reverted to its previous freeholder, who grazed it. When he abandoned farming, the area fell into the category of 'white land' – or space that might be allocated for future housing. Many considered such a step not only uncontroversial but probably helpful to the overall environment, particularly since it would relieve pressure on villages and farmland in the Vale of York. However, Ryedale District Council refused planning consent on the grounds not only that the Backies was an important open area but also because they considered it a significant nature conservation resource.

The developers appealed to the Secretary of State for the Environment to reverse the refusal of planning consent. Although they could not build on the Backies without consent, they nonetheless started to clear away scrub and trees with bulldozers. Local residents were so enraged that several dozen of them physically blocked the path of the bulldozers or installed themselves within them. To defuse the situation the local authority stepped in and issued emergency tree preservation orders; eventually planning consent was given for housing over part of the site, linked to a scheme to tame the landscape near the new houses, but most of the site was left intact. Today the Backies is still much favoured as an informal recreation area. Amid dog roses, ash trees and flowery fields, blackcaps, chiffchaffs, willow warblers and whitethroats pour forth their springtime song, children build houses in the trees and dens in the abandoned ammunition bunkers, teenagers ride mountain bikes and grown-ups stroll with their dogs, while water voles, now scarce in the countryside, busy themselves in the beck.

Landscape ecologists Tony Kendle and Stephen Forbes have explained why stretches of neglected rough space in towns can be rich in wildlife, particularly if the deposited waste has contributed to the underlying rock and soil. They explain:

> Of course most sites are benign even if they are made up of materials such as waste, but unusual conditions can be found and often lead to the most interesting areas. The pH may range from 2 to 12; the nutrient content may range from excess to almost non-existent ... Every textural class and soil classification can be encountered, including some which have never been properly described.[7]

The interface provides similar and often greater wildlife benefits than urban wildscape, and on a much more extensive scale. For plants and animals being driven out of the countryside by modern agricultural methods, interfacial areas provide an obvious first refuge, benefiting not only from their own ecological resources but also from their proximity to that other increasingly important wildlife refuge, gardens. Wildlife habitats often survive in the interface because farming is pursued less intensively, either because the land is fragmented or because the owners are no longer altogether serious about agriculture. Land uses that replace farming in the edgelands, like grazing for horses or the excavation of minerals, scruffy though they may appear, can actually in themselves be better for wildlife than modern industrial agriculture. Horses in the edgelands are reared for casual exercise, not meat production, so those who keep them do not turn their pasture into a ryegrass monoculture as stock farmers usually do; in any case horses, not being cloven-hoofed, would crush reseeded ryegrass. Lakes, like those at Stonar and Molesey Heath, which are allowed to develop after mineral extraction, can be rich in wildlife. Rainwater and groundwater provide the source of their clear water, unpolluted by run-off from agricultural land or rivers into which effluent has been discharged. Having no inflow or outflow such bodies of water contain none of the large bottom-dwelling fish like bream that stir up the mud and silt of most lake bottoms, occluding the water and thus stunting the development of many underwater plants. Young amphibians that would be eaten by such fish flourish in this predator-free environment.

Interfacial sites often enjoy biological diversity partly because they are ignored. Being ignored, they go unmanaged. The clutter of the interface, which would be tidied out of sight by those concerned with creating an acceptable landscape there, often enhances wildlife by creating new niches that wild creatures can exploit. Throw an empty milk crate into a lake and while it may look untidy, fish will swim in and out of it and use it as part of their ecological world. Black redstarts nest in the brickwork of derelict buildings. So while town parks are grassed over for ball games and our national parks overgrazed by sheep, these truer wildernesses are allowed to find their own accommodation with Nature, evolving silently and unhindered.

Art, history, play, utility

Wildlife value is relatively quantifiable; some of the other benefits bestowed by the edgelands are less so. The subversiveness of
interfacial land perhaps explains why children often value it
more than other groups, seeming to find the edgelands a wonderful place to play. Why? This landscape offers an obviously
varied environment, which is often wild, and which has plenty
of places to hide and things to play with. Its dereliction stimu-
lates the imagination. This seems to apply not only to children
but also to some kinds of creative artist. The aura of excitement
that goes with the apparent lawlessness of the edgelands has
been exploited in such films of the 1990s as *Reservoir Dogs, Pulp
Fiction, Seven, Things to Do in Denver When You're Dead, Fargo* and
The Straight Story. The interface is the obvious place for people
to hand over money in brown envelopes or to hide kidnap
victims. Here the final shoot-out can take place against an
ominously incomprehensible industrial backdrop. Strange
pieces of debris of twentieth-century capitalism seem to have
some of the capacity to inspire awe in the same way as the left-
over artefacts of other ages, such as Stonehenge. Like such
ancient relics, the relics of our own age now stand outside the
humdrum landscape of everyday life. These things have not been
designed by some functionary eager to satisfy the bland tastes
authority attributes to the public. They do not greet you like a
country or town park, a high street or any other environment
tailored to the supposed needs of the lowest common denomi-
nator. This is a vaguely menacing frontier land hinting that here
the normal rules governing human behaviour cannot be altogether relied upon.

Yet this living museum of our recent past is also the power-
house of the society of the present. Visit a high street today and
the chances are you will not learn all that much about what
makes the town that contains it tick economically. You would
probably get a much better idea by taking a trip to its interface.
Here you will find the branch offices, industrial works and su-

perstores that probably account for far more turnover than the businesses and shops surviving in the traditional centre. You will also find much of the essential apparatus without which the whole settlement could not function: rubbish tips, electricity sub-stations, sewage works, gas-holders, motorway interchanges and so on.

Neglect and abuse

None of these considerations – wildlife importance, value as an adventure playground for children and imagined landscape for artists, museum of our recent past and cauldron of our economic present – seem to be sufficient to earn the edgelands the respect of those who control our lives. Not only is less care taken over the appearance of new buildings here, but councils also neglect to provide the most basic public facilities they would automatically provide in a town.

Out-of-town shopping is a key component of Westwood on the Isle of Thanet, which is home to huge stores like Sainsbury's Homebase, Tesco Extra, Carpet Right, Halford's and B&Q – more than 37,000 square metres in total – and which now exceeds that found in Margate town centre. Close by, two business parks are springing up. Thanet District Council has chosen not to provide the sort of public facilities that a town centre would offer. Instead, it is relying on the retail operators themselves to provide these: once a further massive shopping development has been built, they will probably include a paved square with benches. Do people who work in the interface in retail complexes and business parks like this not need a park for a stroll at lunchtime? Where, for that matter, are they to eat their lunch? Kick a football with their friends? Or watch their children play? There is, of course, a fear that provision of public facilities in the interface would further hasten the depletion of town centres which is already so pronounced. Yet the absence of basic facilities either encourages people to travel, usually by car, if they have the time, or it leaves them stranded in conditions we would not tolerate elsewhere. Furthermore, the absence of any community space deprives people of the right to live fully. The right of freedom of association and the right to stage public demonstrations can hardly be exercised in edgeland settlements like this where such apparently public space as exists, like superstore car parks, is actually private.

Such neglect by local authorities does not appear to bother central government. In 2000 the Department of the Environment, Transport and the Regions (DETR) published a white paper entitled *Our Towns and Cities: Delivering an Urban Renaissance* and another on the future of the countryside of England called *Our Countryside: The Future*. Both were substantial documents, running to a total of nearly 400 A4 pages between them. The urban white paper makes no mention of the interface, apart from the word 'fringe' appearing in a diagram, yet it is hard to see how urban decline can be stemmed without attention being paid to the role of interfacial areas, which so often suck economic activity out of the centres of towns and cities. The sister document refers to the urban fringe briefly in the context only of the difficulties of farming there.[8]

The main instructions from the DETR to local councils in England on planning on the edge of built-up areas come in the shape of their publication *The Countryside: Environmental Quality and Economic and Social Development: Planning Policy Guidance Note 7*, issued in 2001. This document sees interfacial land as essentially problem land with no intrinsically valuable features: its use is to offer space for functions like waste disposal and informal recreation in order to allow more valuable land to remain untouched.[9] So the interface remains a dumping ground for activities considered unprepossessing and a frontier land in which private sector development rages unchecked by noticeable standards of design. One of the most striking of such activities may soon be wind farming, which is set to boom as power generators are forced to turn to renewables to reduce the consumption of fossil fuels. In the face of complaints that windmills ruin remote moorland landscapes, officials are coming to see interfacial areas as their obvious home, even though they will of course be vastly more intrusive in the intricate environment of the interface. Equally threatening for this landscape, which relishes what other landscapes vomit up and which laughs at current notions of taste, are attempts, also under way, to castrate it by turning it into the sort of landscape considered more desirable. It goes without saying that all three approaches reflect the assumption that the interface has nothing to offer as a landscape in its own right.

Afforestation has emerged as the fashionable mechanism for transforming as much as possible of the interface into something more acceptable to polite society, and hiding as much as possible of the rest of it from view. Vast interfacial areas bordering towns from Bristol, Swindon and London Colney to Bolton, Warrington and Hartlepool, together with much of the edgelands of Romford, Rainham, Nottingham, Manchester, Liverpool and Sunderland, are to be transformed by courtesy of substantial grants into so-called 'community forests'. In England alone (and there are also similar forests in Wales and Scotland) the community forest programme covers an area of more than 1,700 square miles (4,400 kilometres), or over twelve times that of the Isle of Wight. Not all of the land involved is interfacial, but the vast majority is. 'We're transforming the living and working environment of the North East,' proclaims a document which sets out how 101,000 acres (40,900 hectares) will be planted with 20 million trees to make the area more attractive for business and housing and to 'improve the landscape and the image of the North East'.[10] The same document presents pictures of an old industrial landscape which has been bulldozed clear of any upstanding features or natural vegetation alongside the replacement landscape aspired to: a line of trim, brick houses surrounded by a liberal sprinkling of trees and new, screwdriver-assembly factories.

This, it is imagined, is the kind of scene that will lure industrialists and prospective homeowners alike, to fill the grim world of the edgelands with happy workers and laughing families. All over Britain, disused quarries, old industrial land and other varieties of unkempt wasteland are to be turned into something more respectable and legitimate – woodland. The intention for the community forests is not continuous forest: about 30 per cent of the land is to be tree covered, with the remainder 'a rich mosaic of wooded landscapes and land uses including farmland, villages and leisure enterprises, nature areas and public open space'.[11] But the essential idea is that an undesirable landscape is to be turned into something else – green, but ordered rather than wild.

Behind this onslaught on the integrity of the interfacial landscape lies something more than the sort of attitudes to landscape distilled by Lowenthal and Prince. The planning theorists have their own theology of hostility towards it. Not that the edgelands were even noticed by planners until the 1960s, when Alice Coleman, a geographer based at King's College London who was taking part in a land utilisation survey, uncovered the existence of a large amount of fringe land that did not fall neatly into the land-use pattern of either farmscape or townscape.[12] She called this land-type 'the rurban fringe', maintaining that new development had sprawled into the countryside in a way that obscured the distinction between town and country. She noticed that, as a result, farmland had become fragmented and in some cases, abandoned. She considered this state of affairs essentially undesirable, and recommended that the rurban fringe should be reduced or eliminated either by being turned into proper townscape, with neatly rounded-off development or, better still, into productive farmland.

The legacy of Professor Coleman's insights can be found in contemporary attempts to sanitise or otherwise neuter the edgelands up and down the country. Travel on the railway line from Swansea to Carmarthen and you will be struck by two quite distinct landscape forms close to Llanelli, the former steel-working centre. On its western edge is a stretch of wild edge-land: pylons march over a rough sweep of country in which old corrugated iron sheds and gaping factories seem to be sinking in tangled skeins of vetches and meadowsweet, while herons suddenly rise up from what look like forgotten dykes. Sadly, this beguiling kingdom falls within the area of the Llanelli Millennium Coastal Park which, together with another community forest project, is rapidly transforming it into a grassed-over area of flat land and low ridges, the latter sporting ranks of sapling trees in plastic sheaths. The aims of the scheme sound laudable enough (to enhance landscape and wildlife), but although it will generate some new wildlife habitat, mainly through the creation of new wetlands, it will at the same time destroy a unique environment.

Elsewhere change is not only transforming scarce habitat but also eradicating industrial landscape. The gaping factory to the west of Llanelli is now the only surviving relic of the town's nineteenth-century industrial past. Yet in its heyday it produced more tin than anywhere in the world and was known as 'Tinopolis'. Llanelli would not exist on its present scale were it not for the industry that came and went in a brief 150-year period. On the southern edge of the town, for instance, in the Machynys area, a link road and prestigious golf course today provide no hint that here once stood foundries and steelworks, tin works and railway sheds, two villages including workers' dwellings, shops, chapels, a school and an older farmhouse: all have been completely swept away.

The case for planning

The world of town and country planning in Britain has yet to catch up with the intricate and varied needs of the edgelands. Planning authorities pay much less attention to the detailed planning of these areas than they do to that of either town or country. Instead, they continue to allow the interface to be shaped largely by the planning applications that happen to come in, rather than by proactive planning with the use of instruments such as compulsory purchase and town plans to assert a public realm.

When planners do confront the interface they find themselves confronting massive economic pressures. The price differential between agricultural land with no planning permission and that with development value accorded by the granting of planning permission is colossal – between a few thousand pounds per acre and millions of pounds per acre. Developers like interfacial land because it is usually greenfield, and therefore free of existing buildings or noxious waste, and yet it is close to urban areas. Farmers are usually only too pleased to sell. If they wish to carry on farming, it is usually far more sensible to sell a farm in the interface and use the millions of pounds they raise to buy something somewhere else. The task of the developer is made easier by the fact that councils impose fewer conditions in terms of design of buildings and so on than they do in other environments. As a planner who had worked in development control on the edge of Cambridge explained to me, 'It is important to bear in mind hope value: the idea that farmers are waiting with pieces of land that they hope they can sell for development; land where holdings are probably fragmented by roads and so on, so it might be difficult to farm them. There are fields around Cambridge where farmers are waiting to get permission: they feel, "This is my retirement." They see the council vacillating, denying them the opportunity to make a packet. Every time there is a planning debate, say about the local plan, they put forward their plot of land, pleading for it to be built on. So this is farmland in a minimal state, farmed in a desultory way.'

Large-scale development in many of our edgelands took off during the Thatcher years when planning deregulation was fashionable, with central government predisposed towards free-market thinking. Often, one superstore development would trigger an interfacial growth explosion; in the Bristol North fringe, on the Isle of Thanet and in many other areas, planning consent was turned down by the local authority but allowed on appeal by central government. As further applications came in, local authorities felt they had no choice but to grant them, extracting such benefit as they could – an improved road junction here, the guarantee of a special Tesco bus service there – under planning-gain agreements. But they were unable to do anything significant to alter the distinctive buildscape that was emerging – spread out, discontinuous, fragmented and without a clear focus, comprising self-contained units of development and oriented to car-based travel. The absence of any master plan for interfacial developments in the past does not, however, mean that such plans are emerging now.

Several interfacial areas do benefit from the application of Britain's oldest and most popular planning tool – green belt. Interfacial land around Edinburgh, Bristol, Bath, Glasgow, Cambridge, Bournemouth, Manchester, Sheffield and London, among other places, is covered at least partly by green belt designation. No green belts have been designated in Wales, but local authorities are entitled to create them, subject to approval by the Cardiff Assembly. Yet the purpose of green belt, unlike all our other planning designations, from National Park to Conservation Area, is not to protect the land it enshrines. The tool was devised in the 1930s at a time when many towns were exploding in size, but at a time when local councils had yet to acquire development-control machinery. Green belt was therefore invented to protect the integrity of the built-up areas on one side of it and the countryside on the other. Today green belts continue to be designated to prevent the coalescence of neighbouring settlements, to contain sprawl or to preserve the setting and special character of historic towns, without any real concern for the green belt land itself.[13] Their purpose remains the creation of firebreaks between genuinely valued landscapes, not to tease out the essential qualities of the distinctive landscape that develops on urban edges, and to enhance it in its own right.

So neither green belt nor other broad-brush designations used around settlements that lack formally designated green belt recognise the varied character and demands of the interface – an environment at least as varied as townscape, the countryside or the coast. In any case, green belt is easily overridden. Thus around York, the green belt has been rolled out: its boundary has been extended outwards to allow for building around the edge of the built-up area. The total extent of the York green belt is kept more or less the same through the extension of its outer boundary further into surrounding countryside. In other places, from Edinburgh to Bristol, areas within the green belt that local councils were prepared to see developed have been zoned by them as 'land for non-conforming uses' or as 'urban envelopes', and released.

As it happens, the concept of the green belt itself is now in question. The prevention of coalescence seems far less important today than it did in the 1930s, and such prevention flies in the face of arguably the most important planning imperative of the twenty-first century – the need to plan for 'sustainable' development. The existence of green belts has caused development to leapfrog over them to places beyond the green belt. This encourages commuting back to the urban area (while sacrificing countryside that is more untouched than land in the interface).[14] If we are to have planning for sustainable development, it is probably better to have development in the interface or on routes already well served by public transport than on land beyond green belt zones.

Detailed land-use plans could focus specifically on interfacial areas. These plans ought to face up to the lack of public provision in these areas. Should we have public parks in the interface? Should we have cycleways and routes for pedestrians or continue to give these areas over to car use? Should we encourage the development of cinemas, nightclubs and restaurants in the edgelands? Or should we insist that these stay only in town centres, many of which currently suffer from night-time desolation?

The existing buildings of the interface also call out for choices to be made. The giant sheds of the edgelands are not designed for the requirements of sustainable development. Many are built to be temporary, for businesses that expect technology, and thus their requirements, to change within ten years. Artificial light is essential in the sheds, since they are designed without windows. Artificial heating is necessary because no solar panels have been installed. And these buildings require a high degree of car use because they tend to exist outside the public transport network. All of these things could be altered. The massive, gently sloping roofs of the sheds would be ideal for solar panels. Because they are located alongside ring roads and major highways, bus services to them could be provided with relative ease; at the same time car use could be discouraged through a variety of measures, as has already happened in town centres. But if we are to go down this route, real compulsion will be necessary.

What, though, of the lives of the people who live in the interface? Unlike the edge cities of the United States, our edgelands are not often desired as a place in which to live. Nonetheless, they often embrace bits of housing, and outer suburban housing is frequently adjacent. Generally speaking, however, these areas of settlement do not form distinct villages – with a clear edge, on a human scale, with mixed-use development, traffic-calming devices and pedestrian and cycle trails. Should we be converting the housing of the edgelands into mini villages, as the New Urbanism movement of the United States is trying to do in the edge cities of that country? Or should we seek to retain the haphazard, sometimes rather forlorn character of much of our interfacial housing? Also, we assume that housing on the outer edge of our towns and cities should be low-density; visit the outer suburbs of Stockholm, however, and you will see tower blocks cheek-by-jowl with farmland.

Bringing the interface more fully within the realm of town and country planning does, of course, involve a basic contradiction. Much of the special character of interfacial areas arises from the fact that they are not planned and not managed. If the essential feature of the edgelands is that they are untamed, and that they express our own age in being so, then to plan them is to some extent to trample on their essential character. If we conserve the land strewn with wild flowers around Llanelli and prop up the hulks of its once-active industrial works then are we not essentially destroying what makes such an area attractive to start with? If the interface is the only theatre in which the real desires of real people can be expressed, and if we wish to celebrate it and not inhibit it precisely because it provides that theatre, then trying to shape the built environment of the interface so that it becomes like an Italian city centre may appear to rob the interface of the freedom on which its character currently depends.

Although one could argue that it is a contradiction to try to intrude the dead hand of the planner into something whose character is to be free, I nonetheless think that we should. If we do not chronicle and assert the special wildlife and historical value of parts of our edgelands, these will disappear anyway, just as they have at Llanelli and in many other places. If we continue to fail to provide public facilities in the edgelands, then we continue to exclude from them the 30 per cent of households who have no car, and we diminish the lives of those who work and shop in them at present. The way in which we intervene will determine whether we conserve or mutilate these strange spaces. In the context of the edgelands, we need to see the planner not as the shaper of an entire environment but as a handmaiden, who helps along a universe he or she does not seek to control. People could be allowed to walk over the flower-strewn wildernesses of the edgelands without the aid of tarmac paths or signboards. And we could simply refuse to permit further developments of the kind that have caused so much visual damage.

One possibly encouraging sign is the publication, in 1999, of the Countryside Commission's document *Linking Town and Country*, in which the Commission urges local authorities to prepare greenscape programmes for the land within and around the edge of towns and cities.[15] The idea is to create a network of green spaces running out from the heart of towns into the wider countryside. The Commission urges that such programmes should be sensitive to the local environment. However, it does not go so far as to identify the special qualities that councils should look out for in the interface, so it is difficult to see how the initiative will make councils much more sensitive to the special character of the interface than they have been up till now.

Another positive sign has been a decision of the Countryside Agency (which succeeded the Countryside Commission in 1999) to provide funding jointly with local authorities for the appointment of small teams of people charged with seeking to improve the appearance of the urban fringe and extend provision for walkers, riders and cyclists, largely through talking to landowners. The Agency recognises the potential recreational importance of the interface in that it represents land with a rural feel that is also close to where many people live. Thus Molesey Heath, for example, lies within the Lower Mole Valley Countryside Management Project. However, such initiatives, welcome as they are, are paltry in comparison to the support provided to more favoured parts of the planning landscape, such as national parks and areas of outstanding natural beauty. And they do little or nothing to protect the areas involved from undesirable change.

Focus of our age

The neglect and disdain with which the authorities regard the interface at least enable it to continue as the ultimate physical expression of the character of our age, unmediated by the passing tastes of elite groups. In the edgelands, no bureaucrats' diktats force new buildings to mimic the style of the past. So here rises the architecture of our own time in all its majesty. The electricity sub-stations and rubbish tips of the interface perhaps more accurately express the character of our time than Portcullis House or the new Scottish Parliament building. Although so much of the interface induces revulsion rather than affection, this has been the fate of the truly contemporary through the ages. We should not let our sense of shock and mild horror inhibit our appreciation. What goes on inside those giant sheds on the edge of my town? Why was that motorway intersection designed in that way rather than another? How does the local sewage works actually function? Instead of seeing the interface as a kind of hellish landscape to be shunned, we should celebrate it. We should see reservoirs and rubbish tips as sources of fascination not only for the civil engineering and landscaping challenges they present, but for what they can tell us about the way our society is.

Town and country may show us the surface of life with which we feel comfortable, but the interface shows us its broiling depths. If people were encouraged to understand this world more, they might feel less alienated and puzzled by the circumstances of their lives. Many of the activities we have dumped in the interface may be there because we do not esteem them, but perhaps we should esteem them more. We might all be better off if we increased both our understanding of and respect for the apparently mundane yet vital activities that make our society work. In appreciating the present we might enhance our ability to change the future. By embracing the interface we might learn to be more imaginative about new building everywhere and more respectful of unconventional scenes and forgotten industrial landscapes.

We might also come to want to involve ourselves to a far greater extent in some of those essential functions that are enshrined in the edgelands. If schoolchildren were given guided tours of waste disposal facilities they might develop not only respect for the civil engineering, chemical and landscaping challenges that have been overcome in building them, but also a greater concern about the use of packaging and a real interest in participating in debates about how waste is to be processed, by whom and where. They might boycott plastic bags that give rise to disposal problems. Car scrapyards are not only interesting but can evoke new perceptions on contemporary life. Most of us

encounter cars most of the time in a positive light: advertise-
ments make them objects of desire by showing them in pristine
form. When we see how they break, how they crumple and go
rusty, we may find it easier to break free of their stranglehold
over our lives and our environment. The appearance of cars in a
scrapyard reflects their real character: cars are made of metal not
paint, and under their glossy exterior they are jagged, corrodable
objects.

Few people flashing past a railway marshalling yard pause to
think about why rolling stock is stationed there or to consider
how the network of which it forms part actually operates. They
tend instead to gaze over it at the scenery beyond or to cocoon
themselves in a book. We take our unfamiliarity with such stuff
for granted. We have placed understanding and therefore
control of such matters in the hands of others. In the past, when
things seemed simpler, people probably knew much more about
how their world worked. This must have helped make them feel
more in control of their destiny than our alienated citizenry.
Reinvolving people in the mechanics of life might, for example,
help them make more mature judgements over issues such as
safety.

Today, we prefer to celebrate the romantic aura that surrounds traditional activities from the past rather than to grapple
with those of the present. We yearn to live in a medieval
cottage, perhaps a converted forge or farmhouse. Yet when such
a cottage actually functioned as forge or farmhouse it was probably thought of much in the way we think of a sewage works or
a car-breaking yard: noisy, smelly, hot and mundane. Just as
perhaps people in the fifteenth century should have taken more
interest in the forge when it was not old, when it was part of the
lives they were living, so perhaps today we should square up to
the hidden facets of our contemporary lives, especially the functions in which we do not engage directly ourselves.

Kindling an interest

How could we kindle more interest in the edgelands? Black-and-white photography, television, film, sculpture, painting and poetry could contribute much. Some artists have painted features of the interface as well as scruffy landscapes occurring elsewhere, such as shacks and broken-down fencing, but it would be good to see the development of an interest in the interface per se by individual artists or groups of them. It would be even more interesting to see artistic expression of the dynamism that the interface enshrines, rather than simply the decay and redundancy with which artists usually identify it. Such interest could be encouraged through the promotion of exhibitions or competitions by local authorities with a stake in interfacial territory. We could also do with a bit more basic information about the extent and make-up of the interface, of the kind we take for granted where other kinds of landscape are involved. No quantitative information exists about, say, the wildlife populations of the interface. Yet this could be obtained with relative ease. Britain is already divided into kilometre squares whose wildlife is laboriously documented by eager enthusiasts. It would take no great effort to determine which squares contained 50 per cent or more land that could be considered interfacial, and then to extract the wildlife data relevant to these squares. Once we knew what they contained we would be able to speak meaningfully of the wildlife population of the interface.

Other studies could build up a picture of human demography, industrial use, archaeological features and so on. Local civic societies, naturalists' groups, historical societies, town and country planning organisations and educational institutions, including schools, might all be encouraged to participate. Elderly citizens could be invited to describe the changes they have witnessed to their local edgelands during their lifetimes.

Guidebooks and guided walks should open up this new world, just as they do more familiar ones. Each September the Civic Trust takes the lead in setting up a weekend of guided walks to open people's eyes to their own localities. Until now such walks have rarely touched the interface, focusing on town and village centres and the workings of, say, Victorian water mills. Why should such walks not also embrace new business parks or the complex of gas-holders, old factories, allotments, recycling centres, telecommunications masts and industrial sheds, together with the hidden stretches of river and expanses of wild vegetation that often lurk in the interface? Here we can see history in the making, as one use is replaced by another, at least as well as we can in any other part of our environment.

It may seem an uphill task to transform perceptions of a landscape as reviled as that of the edgelands, but history teaches that this may not be as big a challenge as it seems. Those who find it inconceivable that the public's tastes could change so dramatically or that the priorities of town and country planners could be so radically transformed should take heart from changes in taste and policy that have already taken place. As in so many other areas of life, people respond readily when given a lead. Perceptions of landscape are changing all the time. Hitherto reviled landscape types have constantly re-emerged as desirable and fascinating, as the tides of fashion ebb and flow. I have already referred to the suburbs, but there have been plenty of other swings in landscape fashion. In the eighteenth century the idea that stretches of grassland dotted with trees should be viewed as the new top landscape – parkland – must have been quite revolutionary. Moors and mountains were once considered hideous places. Daniel Defoe reflected the accepted tastes of his day when he described Westmorland, with its Lake District mountains, lakes and fells, as 'a country eminent for being the wildest, most barren and frightful of any that I have passed over in England, or even in Wales itself'.[16] Yet in the twentieth century the protection of mountain and moorland became an overriding component of policy and when, in the 1950s, the National Parks Commission came to single out stretches of countryside on which to bestow the highest level of landscape designation – the newly created status of national park – it was to moorland and mountain that they most frequently turned. This reversal of public taste was spearheaded by writers like William Wordsworth and Emily Brontë. When they were living in a remote part of the Lake District and the Yorkshire moors respectively, mountain and moorland were still thought of as places it would be a misfortune to have to visit, let alone inhabit, but their lead helped turn Britain's moors and mountains into magical landscapes of myth.[17]

It is time for the edgelands to get the recognition that Emily Brontë and William Wordsworth brought to the moors and mountains and John Betjeman to the suburbs. They too have their story. It is the more cogent and urgent for being the story of our age.

Lowland landscapes: beyond maximising production?

Ian Hodge and Uwe Latacz-Lohmann

Of all Britain's landscapes, the intensively farmed arable low-lands are probably the most shaped by the requirements of agri-cultural production. This in turn is influenced by the economic situation, by technology and by public policy working through the agricultural decision-making process. Changes in these fac-tors have affected the landscape in the past and will continue to do so in the future.

The appearance of the landscape in lowland areas depends very much on the extent of land that is in arable production. Figure 3 shows the total area of tillage, representing cropped area and fallow land, in England and Wales from the middle of the nineteenth century up to the present day. This demonstrates the long and steady decline through the second half of the nine-teenth century, when tillage represented almost 50 per cent of the agricultural area, continuing into the twentieth century, with the First World War as a temporary peak, prior to an accel-eration of the decline through the 1920s and 30s, when tillage fell to about 30 per cent of the agricultural area. The Second World War saw a dramatic increase in the area of tillage. This was partially lost in the following decade, but was then steadily re-gained from the mid-1960s to the mid-1980s. Most recently we have witnessed another decline.

In reality the impact of these changes on the appearance of

Figure 3 **The area of tillage in England and Wales**
(in thousand hectares)

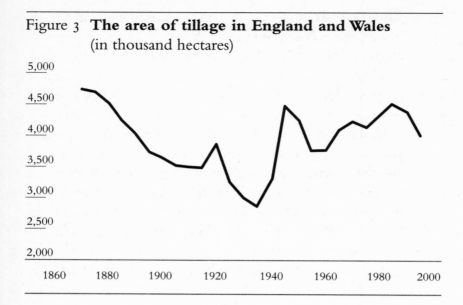

the countryside may well have been even more notable than implied by the figures – the area of tillage was considerably more concentrated in the eastern counties in the mid-1960s than it had been in the late nineteenth century.

The drive for production

During the period before the Second World War agricultural depression and weak planning policies had resulted in substantial areas of abandoned agricultural land. Derelict land was a common sight, with the 'thousands of lost acres that defaced Essex at the outbreak of the war in 1939'.[1] During the war considerable efforts, directed by the County War Agricultural Executive Committees, were devoted to bringing this land back into production. It has been estimated that nearly 8,500 acres were 'reclaimed from bushes' by the Essex War Agricultural Executive Committee. This process of reclamation was seen as generally improving the landscape. A. G. Tansley, a leading ecologist, commented that:

The great extension of agriculture during the war has
not on the whole diminished the beauty of the
countryside – rather the contrary is true. The increase in
arable agriculture, with the corresponding disappearance
of 'permanent grass', has distinctly improved the aspect
of the country in many places, because it has replaced
monotonous and often practically deserted and derelict
grass-fields by the varied and stimulating activity
associated with the plough-land. In places no doubt,
heath has been destroyed and fenland drained and
ploughed, and some of these changes have been deeply
regretted by the naturalist and the lover of nature. But
the total loss has not been very severe, and is offset by the
gain in the agricultural area. It is scarcely probable that
the extension of agriculture will go much further, for the
limits of profitable agricultural land must have been
reached in most places.[2]

Interestingly, Tansley goes on to express concerns about the
potential increase in the extent of conifer plantations. Areas of
special scenery and native vegetation should be scheduled and
protected from 'development'. In other areas, which had already
been considerably influenced by human activity, all that was required was to continue that activity.

The period of change after the Second World War can be
seen as one of modernisation, with the adoption of new technologies introducing a considerable simplification and standard-
isation in agricultural systems. This is not to say that the new
technologies did not involve new skills and abilities, but the
numbers of different enterprises on farms declined and the variations in systems between localities were lost in the pursuit of
higher production and lower costs.

Prices were supported at levels above those in world markets
and guaranteed through agricultural policy. Combined with the
availability of new technology, this stimulated greater intensity
of production, higher use of chemical inputs, investments in the

agricultural 'improvement' of the land through land drainage, reseeding, removal of natural features and rationalisation of field patterns. The substitution of machinery for labour (and animal power) increased the scope for economies of scale, both in terms of enterprises within holdings and in terms of the holdings themselves. Bryn Green discusses the ecological impact of these agricultural changes.

The impacts of land ownership and farm structure

The post-war period saw a continuing growth in the predominance of owner-occupation. Some indication of this is evident in the statistics collected by the former Ministry of Agriculture, Fisheries and Food (MAFF), though there is some uncertainty as to their interpretation in practice. In 1914 the percentage of agricultural land recorded as rented reached a peak at nearly 90 per cent. This fell immediately following the First World War, to 82 per cent in 1922, 62 per cent in 1950, 50 per cent in 1960 and 48 per cent in 2000.

It is widely believed that small 'family' farms are better for the environment than alternative structural arrangements. But this leads us into a number of difficulties, some of which come down to definition, and others that result from lack of evidence. The changes that are taking place in the agricultural labour force do not indicate any general decline in family farming. In several parts of the country, particularly in the east and south of England, where the proportion of family labour is already relatively low, family farming is in fact on the increase. Rather the explanation lies in the more rapid decline of hired labour. In the more pastoral northern and western parts of the country, the proportion of family labour is declining, probably because of a decline in the numbers of spouses working on farms. Nationally, however, the proportion of family labour remained pretty constant throughout the 1990s.

In practice it is difficult to find any clear evidence that small farms are better for the environment than large ones. Generally, the intensity of production found on farms is similar across all farm sizes, arising from the fact that all farmers are adopting very similar technology. There is, however, likely to be more variation amongst small farms, as a reflection of the diversity of goals and values of farmers. A few small farms, no doubt, remain as relics of a past age. Some older farmers may not have gone through the process of modernisation undertaken by the great majority, and so retain more labour-intensive, mixed farming systems, maintaining hedgerows and rough land, and fail to exploit the full productive potential of their holdings. It is more likely that such farmers will apply their inputs with less care and precision, fail to adopt modern waste-treatment and pollution controls, and not have the capacity to undertake positive landscape management practices.

In order to survive in times of low incomes, small farmers who are financially dependent on farming will be under considerable pressure to extract the maximum output from the resources available to them, intensively farming every hectare. Therefore, it is those farmers who can afford to forgo some of their agricultural income in order to provide space for wildlife, and who have spare labour for landscape management, who do most to conserve the landscape. These tend to be the hobby farmers and managers of larger businesses.

The influence of land tenure on the environment is similarly difficult to predict. Owner-occupiers have a long-term stake in their holding, especially where the following generation is expected to take over. But tenants too may have a long-term interest, with perhaps a similar expectation of family succession. Much will depend on the nature and outlook of the landlord. We might expect differences between landlords who give an emphasis to the environment, such as the National Trust, which undertakes detailed ecological assessments of its farmland and includes conservation covenants in its tenancy agreements, and those who place a primary emphasis on the financial return. There will be a similar variety of perspectives amongst personal landlords. Land ownership and farm structure are but two types of influence among many, and probably not the most important. The prices that farmers pay for their inputs and receive for their products, the laws regulating the way in which they farm, the policies influencing their activities, and the opinions and judgements of other farmers may all be of greater consequence in conservation terms.

Reforming the CAP

The period between 1984 and 1992 saw the addition of a range of policy mechanisms designed to restrain the levels of agricultural production and their associated budgetary costs. Most prominent amongst these was the introduction of dairy quotas in 1984. Measures were also introduced into the arable regimes, but these were of a rather small and technical nature and made little fundamental difference. Thus, co-responsibility levies and rules about the standards of grain delivered into intervention were introduced in the 1980s. These effectively reduced the prices received by farmers, but in rather complex ways. 'Set-aside', a familiar feature of United States agricultural policy, was also introduced into the European scheme on a small-scale, voluntary basis in 1988. But this modest tinkering was insufficient to resolve the more fundamental problems of agricultural policy.

A number of factors combined to jump-start some rather more radical changes to the Common Agricultural Policy. The failure of previous adjustments to resolve the underlying internal budgetary problems meant that there were continuing internal pressures for change. But almost certainly more persuasive were the pressures emerging from international trade negotiations. With the initiation of the Uruguay round of talks under the General Agreement on Tariffs and Trade in 1986, agriculture was brought into trade negotiations for the first time. These negotiations encouraged changes in the ways in which support was being provided. The CAP reforms introduced under Com-

missioner MacSharry in 1992 did more to ease pressures on production by introducing a whole new style and vocabulary into
arable production policy. The level of price support was to be
reduced towards world-market levels and set-aside was introduced on a broader basis. In return, farmers were to receive payments based on the area of land in production.

These reforms changed the incentives for farmers when
making decisions about their production activities. While the
total budgetary cost of support changed little, indeed it was expected to increase somewhat, there would be important changes
in the way in which it was delivered. The CAP reforms introduced limits on the total area of arable land that could receive
support payments, reduced the level of support that farmers received for production at the margin, and introduced a requirement on most farmers to set aside some of their land.

The idea of set-aside, paying farmers not to grow crops, at
first glance looks to be born of a rather perverse logic. In fact the
logic is simple. CAP support creates incentives for farmers to
produce products beyond the level of domestic consumption
and at a price at which they cannot be sold competitively on
world markets. Excess produce thus has to be stored and sold
with a subsidy on the world market. If the cost of disposing of
the excess is greater than the cost of paying farmers not to
produce it in the first place, then set-aside makes sense. But only,
of course, if reduction of the initial incentives is not considered
to be an option.

Despite the central role of set-aside the approach has been
the subject of substantial criticism. For instance, the Minister of
Agriculture, Fisheries and Food's CAP Review Group commented in 1995:

> Restrictions on production cause serious distortions in
> the operation of the market, restricting the ability of
> more efficient farmers to gain at the expense of the
> less efficient, increasing bureaucracy and creating scope
> for evasion and fraud ... An excessively high level of

set-aside would adversely affect the rural economy and the appearance of farms as well as giving the erroneous impression that we have 'surplus' agricultural land.[3]

A price reduction would be the preferred way of restraining production, as this would promote a much greater adjustment in production patterns. Market incentives would encourage low-cost producers to expand their areas while forcing high-cost producers out of business. Thus the overall cost of production would be reduced. The further cost of set-aside in the form of the payments made to farmers by government could also be avoided. As this shows, a substantial cost is incurred from ruling out the possibility of a more fundamental adjustment.

Attitudes towards set-aside and its impact on the environment have altered with experience. While farmers were initially against it, they now widely accept the policy and indeed most would be reluctant to see it go. Farmers have effectively incorporated set-aside into their crop rotations with the result that they see it as having little or no overall impact on their businesses. In terms of environmental impacts it seems that, on balance, set-aside is beneficial. The initial concerns that set-aside would have detrimental effects on the landscape appear to be less often expressed. At the same time expectations that set-aside would lead to consistent and significant reductions in chemical emissions seem not to have been borne out. Breeding birds are amongst the main beneficiaries of set-aside and the benefits are greater within a relatively intensive arable landscape. For most other environmental aspects the buffer supplied by set-aside is probably more important than its direct impact.

The pressure for further policy change

The process of reforming the CAP has a long way to go. The 'Agenda 2000' package of measures, agreed in 1999, has taken things a little further but was far from radical in almost all respects. The one exception to this may come to be seen as the more formal recognition of rural development as the second key component or 'second pillar' of the CAP. But at this stage few resources have been made available to give it much substance.

Beyond the continuing internal debate about the extent to which production measures dominate CAP expenditure and the extent to which CAP expenditure itself continues to dominate total EU spending, two external forces are likely to motivate further reform. The prospect of admitting up to twelve Central and Eastern European Countries (CEEC) into the EU presents particular challenges to the continuation of the CAP. The principle of common prices is central to the purposes of the EU. At present, agricultural prices in the CEEC tend to be well below those within the EU. The cost of simply extending current mechanisms to these twelve would be excessive, particularly in view of the relatively untapped productive capacities of many of these countries. Yet it could be argued that the direct payments made to farmers under the CAP should equally be extended to farmers in the CEEC. This would represent a substantial burden on European taxpayers. In fact there is a complication here. The payments introduced under the MacSharry reforms were expressed as compensation for the reductions in support prices. But the CEEC farms have not experienced price reductions, rather they are looking for price increases. So what is the logic to justify direct payments?

A second external factor arises in the form of the renewed negotiations on international trade through the World Trade Organisation. Here, agricultural exporting countries, especially those represented in the Cairns Group,[4] are looking for further reductions in the levels of agricultural support. The preparations for these negotiations have spawned a new area of debate. Concerns that liberalisation of agricultural policy would promote a loss of values from the countryside have prompted a search for a more careful characterisation of the wider variety of benefits that are seen as arising from agricultural production. And this is now widely recognised in terms of the 'multi-functionality' of

agriculture.[5] Agricultural production is seen as having the potential to provide landscapes, biodiversity, support for rural
economies and communities, and food security. It is likely that
any agreement will require countries further to reduce any
support linked to production and to target remaining payments
from government more precisely on specific public benefits.

The future for the lowland landscape

The response to lower prices
Economic theory predicts that farmers will reduce the intensity
of their production in response to lower output prices, although
many argue that in the face of lower prices farmers will simply
farm more intensively in order to keep up their incomes.
However, we should note that farmers' leaders did not in fact
advocate a price cut during the drive to increase production in
the immediate post-war period and that, in practice, if farmers
were able to increase income by farming more intensively, we
might expect that they would have done so already. Given that
farmers cannot be expected to ignore opportunities for improving their incomes where they are available, we can be sure that
for the majority and over the longer term, lower prices will lead
to lower intensity of production and lower output.

Within this general tendency, we can expect adjustment in
two respects. First, the lowering of the intensity of production
will take place across the board. A lower output price means a
lower rate of return on inputs and so farmers will tend to use
fewer of them. This will apply to all farmers, who will look for
ways of reducing their costs. But second, we should also expect
a complete withdrawal of arable production from marginal production areas. Historically, discussion has often been concentrated on the movement of cereal production up and down the
hills, as production on poorer-quality land became more or less
profitable. In the hills this might mean a complete cessation of
agricultural activities altogether. The same argument would also

apply to land less profitably cultivated in the lowlands, wet areas or small awkward fields. Thus, at lower prices we may see land abandoned as it was during the 1930s, with similar implications for the landscape. From a historical perspective some retreat from the current extent of production may not be undesirable – the present intensity of production is often a cause of environmental damage – but it is difficult to judge at what point it may be deemed to have gone 'too far'.

The impact of withdrawal of arable production from the marginal arable areas such as the South Downs will be of particular significance for the environment. It is possible that a new pattern of cereals, livestock and input prices might promote the desired pattern of extensive grazing required to achieve the favoured landscape but it is more likely either that land will be taken out of production altogether and abandoned, or that it will remain in intensive arable production. In either case, some agri-environmental incentives would be needed to promote the desired pattern and intensity of land uses.

Different approaches

We should expect different responses from farmers to lower prices, depending on their particular circumstances. Not all farmers regard income from agricultural production as their major motivating factor and there will be no single approach for future arable production. Land managers will have a much wider variety of objectives and purposes than has been the case over the past 50 years, but generally we might distinguish between those who require a commercial return and those who view land primarily for its direct environmental qualities. The former may be further split between those who undertake the production of bulk agricultural commodities and those who produce more differentiated products for niche and specialist markets, and adopt a variety of 'alternative' agricultural production systems.

The commercial bulk producers will continue to dominate the land area and will be under continuing pressures to minimise costs. This assumes that such production can be profitable within the UK at world prices. We believe that it can be, at least in the main cereal areas in the east and south-east. But it will require a sophisticated, high-technology, large-scale approach that we might characterise as a 'lean-burn' solution, as with car engines that are designed to minimise pollution by using as little fuel as possible. Information will increasingly become the key input, permitting the carefully targeted control of crop production at the minimum cost. Gone will be the days of casual prophylactic treatments, rather each application of fertiliser and pesticide will be carefully assessed against site-specific conditions and yield risks.

The pressures to adopt new technologies wherever they can contribute to competitiveness will be intense. This might present particular concerns where such technologies include the use of genetic modification (GM). If farmers in other countries are adopting GM as a means of increasing efficiency and cutting costs, those who fail to do so may not be able to survive. We may then anticipate considerable tension between those wishing to adopt GM technologies and those whose livelihoods are threatened by this, perhaps organic farmers, and possibly the general public. We can expect some gains in the lean-burn solution for the landscape and environment generally in the form of lower rates of chemical use, which should reduce chemical and nutrient losses and the consequent damage to the environment. But at the same time there may be little relief from the coverage of arable production. Production will decline, but more because the land is farmed less intensively than because land is taken out of production entirely. Incentives will remain to continue with production wherever it is profitable.

We may think of the alternative commercial approach, producing more differentiated products, in terms of 'boutique' production. Farmers will find different ways of adding value to their production. The most obvious example is organic farming, but there are other specialist niche markets for pharmaceuticals, wild flower mixes, or locally branded products. The areas involved will be relatively small but they will have local significance for the landscape.

The nature and extent of the non-commercial approaches are harder to predict. Much will depend on land prices and the planning system. We can expect lower returns to farming and so lower agricultural land prices, giving non-farmers greater scope to buy up areas of land for residential and recreational uses. These will include both private individuals and public and private organisations, with land taken up for extended gardens, ponies, nature conservation and access. Some may wish to 'play' at farming but without the discipline of having to make a living. Much of this farmland may well become unkempt and neglected. It remains to be seen whether this will come to be regarded as a new problem for the landscape.

The new structure of farming

We can expect both the commercial and non-commercial developments to lead to changes in the structure of the agricultural industry. Operations will have to be conducted on a large scale in order to be profitable. But whether or not the land will come into large ownership units could be increasingly irrelevant. Farming companies will gain control over land under a variety of arrangements: contracting, renting and owning. The familiar association between land ownership, occupation and control is already showing signs of weakening.

We generally think of the business of farming in terms of a single farming unit, where the occupation and often the ownership of the land is assumed to coincide with the core of the business enterprise. The decision-maker on the agricultural holding, the farmer, determines the way in which the land is managed and it is this, in the context of the current level of output prices, input prices and technology, that establishes the income generated from the core agricultural business. Farm income is a major determinant of the farmer's and the household's level of welfare and establishes both the capacity and incentive to keep the land in agricultural use.

The arrangements by which land is farmed are already becoming much more complex. Legislation introduced in 1995 on agricultural tenancies has given far more flexibility in the choice of arrangements between landowners and occupiers. There have been changes in the legal status of farming organisations, whether as individuals, partnerships or companies, and a significant number of farmers are contracting out a major part of the operation of their holding. Whether this represents an ageing population of farmers who are effectively retiring from an active role, or an increase in the number of 'farmers' who are primarily occupied away from the holding on non-agricultural activities but who choose to maintain an involvement with agricultural management, is not clear. In general terms, however, these changes represent a weakening of the relationship between the occupation of land and the definition of the core businesses within the agricultural sector.

We can expect these trends to continue and even to accelerate under a liberalised policy regime. The comfortable concepts of family farming, of the local farmer, or of the farm as a specific parcel of land passed down through generations will have even less relevance in the years to come. The individual aspects of traditional farming – owning the land, living on the land, cultivating the land – will be separated out and parcelled up for individual consumption. The actual business of production will be in the hands of large, highly technical organisations with little or no local association. It is their activities that will ultimately determine the shape of the landscape over most of the arable area.

Policies for the lowland landscape

The scenario described so far does present a rather lifeless picture of the future lowland landscape, with most of the area efficiently managed in technical terms but without much positive provision for wildlife or landscape. This will do little to promote the resurgence of wildlife or the enhancement of the landscape that we aspire to, an objective whose importance has been emphasised by the inclusion by the government of the population of wild birds as a headline indicator of sustainable development, in areas where the numbers of farmland birds are seen to be in decline.[6] We do currently have in place a number of agri-environmental policies that concentrate on special areas, as discussed by Bryn Green and Philip Lowe, and these are of consequence in the lowlands as well as in the uplands, but primarily apply to more marginal contexts, such as wetlands or downlands. They will do relatively little to conserve the generality of the landscape of the lowlands. We therefore suggest two other policy models that may have a role to play in meeting this objective.

Environmental compensation

Continuous arable production, however lean it may be, tends to leave little space for wildlife. A report in the late 1990s[7] argued that the major impact of pesticides on wildlife was less through the chemicals that harmed the wildlife directly than by the removal of the pests and other species that constituted the basic food sources for birds and mammals. One of the recognised benefits of set-aside is that it obliges farmers to make space for wildlife across even the most intensively farmed areas of the country. Intensive arable systems may be seen as depleting the natural capital of the countryside. Thus, the idea of environmental compensation suggests that in order to be sustainable, we ought to put something back, so as to maintain the natural capital.

This might be done through a modified approach to set-aside. It would be voluntary but competitive. Farmers would tender bids to take areas of land out of intensive production and manage them for wildlife. The government would accept bids in order to maximise the environmental benefit that could be obtained from the available funds. The schemes offered by farmers would be designed to meet public objectives: habitat for wildlife, buffer zones beside watercourses and nature reserves, public access, and so on. Farmers might wish to offer set-aside on a rotational basis, moving the set-aside area around their farms, and this would be acceptable if there were accompanying suitable management practices for wildlife. Rather than concentrating on areas where there are already high environmental values, such as in the Environmentally Sensitive Areas, bids would be favoured where they offered relief from the most intensive agricultural production and where the pressures of human population were highest. Thus environmental compensation could be concentrated within the most intensively farmed and pressured lowland areas. Contracts could be offered to farmers over varying time periods, thus maintaining flexibility to increase levels of food production should circumstances demand it. If perhaps 10 per cent of the arable area were to be held in this new form of set-aside, it could make an appreciable difference to the quality of landscape and wildlife in the most intensively farmed parts of the country.

Arable stewardship

In 2002 the government announced the extension of the Arable Stewardship Scheme. This scheme aims to enhance biodiversity and benefit species associated with arable farmland that are in decline or scarce. Arable Stewardship was introduced as a pilot scheme in two areas, East Anglia and the West Midlands, with the objective of recreating and enhancing wildlife habitats in arable areas. Within these two areas farmers were offered a range of management options designed to create food sources and suitable habitat conditions for a variety of birds, insects, mammals and plants. The pilot scheme has shown that these changes in farming practice can be effective in promoting wildlife. As a result, new arable stewardship options are now included in the Countryside Stewardship Scheme available to arable farmers throughout the country. There are three categories of options: *wildlife mixtures* – grown on strips or blocks of land to provide seeds, pollen and nectar for birds and insects; *overwintered stubbles* – land left unploughed over winter following a cereal crop; and *conservation headlands* – land surrounding cereal crops that is not treated with insecticides or herbicides.

We see these two approaches towards environmental policy for arable production as complementary. Environmental compensation should be applied with a broad coverage and a light touch, with relatively low administration costs. Arable stewardship is more intensive and demanding in environmental conservation terms. It would be more experimental, exploring alternative production techniques and providing information to guide the development of both the environmental compensation approach and that of arable production systems more generally. Together these schemes should offer an alternative to the more commercially oriented systems that can be expected from market incentives.

What impact will these changes have on economic activity in lowland regions? The answer is almost certainly 'not much'. Particularly in lowland areas, agriculture represents a very small proportion of the total level of economic activity, typically less than 5 per cent, even in rural areas. While agricultural employment has declined, most lowland areas have gained population through migration associated with commuting, retirement and a growth of non-agricultural employment. Declining opportunities in agriculture are obviously critical to the individuals affected, but impacts on the local economy will not be noticeable. The wider significance lies in the consequences for the landscape.

Conclusions

Farming in lowland landscapes has been substantially driven by the profitability of bulk agricultural production. And this in turn has been driven by the supports offered to farmers through the Common Agricultural Policy. In the future incentives will be more complicated: future public funding will be challenged by other countries through the World Trade Organisation, which wishes to see an end to agricultural support, and any government payments made will have to be seen to be in support of environmental benefits rather than in support of production.

We anticipate that this will mean lower prices for agricultural products than those that have generally been enjoyed over the post-war period. This does not necessarily mean that prices need to fall much beyond their present levels – much will depend on the ways in which world markets develop. Predictions here are an uncertain art and take us well beyond our present remit but we believe that arable production can be commercially viable in much of the lowlands at current market prices. The present crisis in farm incomes indicates, however, that this will only be achieved through the pain of significant adjustment.

Arable landscapes are essentially working landscapes and farmers must be able to respond to the requirements of commercial production. They will need to adopt new systems and technologies if they are to achieve competitiveness in an international context. We think it unlikely that the agricultural production patterns that will follow from a more liberalised agricultural market will generate the landscape demanded by the public – policy measures will still be needed in order to deliver the environmental benefits. In addition, such policies must be flexible enough to develop as more is learnt about their costs and environmental effects. The idea of combining an efficient means of producing *environmental* outcomes with commercial agriculture is a novel concept and one about which there is much to learn. It takes us beyond maximising production.

After foot-and-mouth: farming and the new rural economy

Philip Lowe

The year 2001 must stand out as a particularly *annus horribilis* for the UK countryside. With farm incomes at their lowest levels for a generation, foot-and-mouth disease struck. The resultant epidemic and the response it induced were on an unprecedented scale.

The disease devastated the livestock industry across much of the west and north of the country, but also cut a swathe through the rural economy. Recovery will take several years. But getting back to normal may be neither possible nor desirable – the spread of the epidemic revealed unacceptable practices. More generally, contemporary agriculture has become so crisis-prone as to call into doubt its entire underlying strategy.

A distinctive feature of the foot-and-mouth crisis, compared with previous farming-and-food crises, was the way in which it pulled down the rural economy. This calls into question the conventional wisdom that sees farm-based diversification as a panacea for reducing the dependency of farm families and rural areas on agriculture. Furthermore, public and political responses to the epidemic have revealed the profound ambivalence in British society towards the contemporary rural economy and the lack of understanding of how much it has changed in recent decades. The hope must be that in shattering some illusions and highlighting or questioning certain

practices and basic assumptions, the foot-and-mouth crisis ·
represents an opportunity to rebuild agriculture and damaged
rural economies on a sounder and more secure footing. The aftermath of the foot-and-mouth crisis presents both the necessity and the opportunity for renewal.

The conduct of the foot-and-mouth crisis

The outbreak of foot-and-mouth disease (FMD) in the UK was
confirmed on 20 February 2001. Within three days MAFF had
banned the movement of farm animals and a policy of slaugh-
tering all animals on infected farms was pursued. In mid-March,
with the disease seemingly out of control, the slaughter policy
was extended to include livestock on all farms neighbouring
confirmed cases. The immediate concern in tackling the disease
was the costs and loss of exports to the UK livestock industry –
an industry that had suffered four years of depressed farm
incomes and that was struggling to recover from the aftermath
of the BSE and swine fever crises.

 To avoid any risks of spreading the disease, many rural organisations quickly cancelled sporting and recreational events,
and the general public was discouraged from journeying into
the countryside. The government closed the extensive network
of public footpaths in rural areas, National Park authorities
asked people to stay away and major visitor attractions shut
down. After a few weeks, however, it became evident that businesses dependent on rural tourism and leisure activities were
also beginning to suffer badly, as people stayed away from the
countryside. Having initially been viewed as an issue of animal
health, i.e. a problem of the agricultural economy, the FMD
outbreak soon precipitated a crisis of the rural economy, as the
discouragement of visitors to the countryside compromised a
much wider range of rural business activities beyond farming.

 FMD compounded many of the economic pressures being
experienced by the agricultural industry in the UK. In what was

the largest foot-and-mouth outbreak ever to have occurred in a developed country, 6 million animals (mostly sheep) were slaughtered, nearly 14,000 farms were left with no livestock and over 130,000 farms (out of a total 180,000) suffered movement restrictions. There were knock-on effects on upstream (animal feed) and downstream (livestock markets and hauliers) activities. However, the crisis also resulted in often severe financial losses for other sectors, including rural shops, pubs, restaurants, hotels, guesthouses and visitor attractions. The resultant economic hardship consolidated and rendered more visible a new economic constituency in rural areas, the rural tourism industry, which actually contributes more to GDP and employs more people than does agriculture. In some areas, hoteliers organised protests to draw attention to their plight, and some were critical of the role and treatment of the agricultural industry. Farmers were compensated for the slaughter of their animals but tourism businesses received no compensation for their own losses.

Beyond farming and tourism, a third tier of businesses suffered. A telephone survey in early April of rural firms in the north-east found that 40 per cent were adversely affected, with 28 per cent suffering losses of more than 10 per cent of turnover. The worst-affected sectors were hospitality, recreation and culture, transport, and land-based professionals and businesses (vets, tree surgeons, garden centres, etc.): in each of these sectors a majority of rural firms were suffering.[1]

The fact that the news media were dominated by coverage of the crisis for several months raised public and political awareness of agricultural and rural development issues. Many commentators used the crisis to call into question the approach to agricultural support currently embodied in the CAP and the prime minister signalled the need, once the crisis was over, for a fundamental rethink of the agricultural industry and its role in the rural economy.

By highlighting or questioning certain practices and basic assumptions, the FMD crisis represents an opportunity to re-establish certain activities on a sounder footing. For example, the extensive animal movements implicated in the rapid and widespread dispersal of the disease seem problematic not only from a biosecurity point of view but also from the perspective of animal welfare and sustainable development. Likewise, the denuding of large parts of upland Britain of much of its grazing herds has also created a one-off opportunity to reduce overstocking and overgrazing in vulnerable areas, and a chance for a radical overhaul of hill-farming subsidies. Finally, it is clear that the worst-affected areas, for example, in the north and south-west of England, need special targeted rural recovery programmes that should provide scope for trying out new approaches to rural development, with a view to building more robust local economies.

Issues for public policy

Animal movement

In the three weeks before the foot-and-mouth outbreak was discovered, Cabinet Office figures show that about 2 million sheep were moved about the country.[2] Movement of replacement breeding ewes and store lambs for finishing from the breeding flocks in the uplands to the lowland finishing flocks is an essential part of the sheep-industry structure, designed to make best use of natural resources. However, the outbreak highlighted the opportunistic role of sheep dealers in this process. Animals were being bought and resold through markets in different regions of the UK over very short periods of time, with some animals going through a succession of different markets and staying briefly on a succession of different farms. Sheep not recognised as infectious therefore came into contact with large numbers of previously uninfected animals over wide areas. In consequence, the first 22 confirmed cases of the disease were in 10 different counties spread across the United Kingdom. The problem was compounded by 'unofficial' dealing at the markets of animals which had never been registered in the official sale records and which therefore were not readily traceable as dangerous disease contacts.

Much of this movement was due to dealers seeking marginal market gains. It would be desirable, from the point of view of both biosecurity and animal welfare, greatly to curtail the scale of live animal movements. Strict constraints already apply to pig movements between farms and these should be extended to other species, including the requirement for a specified quarantine period before newly arrived stock can be moved off a farm. Similarly, more reliable methods are needed to ensure the traceability of livestock movements − electronic implants are developing rapidly and these have the potential to automate location-recording of all individual animals.

A question mark must surely hang over the traditional live animal auction markets. Pressures to phase out the markets must be anticipated, given the central role they played in the spread of the disease, as well as the animal welfare objections to their operations. Controls over live-animal dealers would affect the viability of these markets and some of them may well not survive the crisis. Electronic markets are already in existence and IT developments make it quite feasible for animals to pass through virtual auctions, followed by direct transfer from farm to farm. However, markets are an important element of rural life and they are often at the heart of local towns. They should be encouraged to move into other fields, such as running electronic auctions, providing commercial services for rural businesses and hosting farmers' markets.

More generally, a return to more localised meat production and processing chains would be desirable. A significant recent development has been a major reduction in abattoir numbers, entailing increased journey distances and travelling time for slaughter stock. The closure of local abattoirs has been primarily a result of the growing overhead cost of implementing food hygiene regulations for small businesses. Although the risk of disease spread associated with travel of animals for slaughter is much less than that consequent on movements between farms, it would still be desirable to see more local abattoirs once again, in order to reduce live animal movements. Ultimately, to achieve this objective may require legislation limiting the distance or time that slaughter stock are allowed to travel.

Extensive and intensive farming

Media commentators were quick to assume that FMD was another adverse consequence of 'intensive farming'. This was presumably based on the views that animals kept at high densities are more stressed and susceptible to infection, that the hygienic conditions are poorer and disease challenge greater, and that disease could spread rapidly between adjacent groups. Whilst the first and third points are generally true (though not necessarily the second one) they did not play any significant role in the 2001 outbreak, which occurred primarily in the most extensive livestock production systems. Even so, the refocusing of public concern on 'intensive farming' will increase pressure for the government to reconsider agricultural methods and further promote concepts of sustainability and organic production (as well as tightening up on the rules and procedures to prevent infected meat entering the UK).

At the regional level, a higher density of livestock farms obviously poses greater risk of farm-to-farm spread once a disease outbreak has been initiated. Fuelled by renewed interest in biosecurity arguments, in addition to existing environmental considerations, both farm and regional livestock density standards need to be considered. Such standards already operate in a number of European countries and in UK organic production systems.

In areas of significant animal concentrations, the slaughter of large numbers of animals on contiguous farms could have been prevented by vaccination. The government was prepared to take this step at the height of the crisis, even though it could have prolonged restrictions on farm exports, but did not do so in the face of opposition from the National Farmers' Union. The scale of the subsequent slaughter not only aroused public dismay and disgust but also created short-term and possibly long-term environmental problems through the hasty establishment of incineration and burial sites across the country. The disposal of animal carcasses was on an unprecedented scale and with uncertain consequences for the contamination of the land, air and water sources in the localities where it occurred. The use of vaccination to regulate future outbreaks of FMD must be reconsidered in the light of the problems for the environment, for tourism and for the rural economy that arose by not using it in the 2001 outbreak.

The culling of large numbers of hill livestock as a result of FMD has raised questions about how the restocking of the hills might best be achieved. Returning to 'business as usual' may not be possible after the crisis, even if it were desirable. A survey of 128 affected farmers carried out in 2001 found that 6 per cent planned to quit agriculture altogether and a further 36 per cent did not intend fully to restock their holdings once the crisis was finally over.[3] This could provide opportunities to restock on a more sustainable basis. Past support measures for upland farming have encouraged overstocking, with the resultant overgrazing having detrimental consequences for landscape and nature conservation in vulnerable areas, leading for example to the loss of heather moorland. Overgrazing may result from too many animals but also from reduced management. An accelerated exodus from upland farming could lead to the amalgamation of holdings with a switch to 'ranching' of the hills and the withdrawal of labour-intensive practices, such as shepherding, with consequent environmental losses through both overgrazing and undergrazing (leading, for example, to bracken encroachment).

This will be a particular issue where hefted flocks have been culled out. To re-establish such flocks will necessitate intensive shepherding over several years. It is unlikely that there will be enough experienced shepherds to re-establish all hefted flocks. Alternative land-use options should be considered for some upland areas, including the managed retreat of grazing and the re-establishment of natural woodlands, to create new landscapes and to bring new activities to isolated areas.

Implications for the rural economy

A crisis such as foot-and-mouth challenges fundamental assumptions by revealing underlying realities. What this crisis has revealed above all is how much the countryside has changed in recent years and how out-of-date are official and public conceptions. The last major FMD outbreak in the UK was in 1967. Significantly, both the major Committee of Inquiry and the economic analysis of the 1967 outbreak considered solely its impact on the agricultural sector.[4] In those days the countryside was largely a farming domain. Much has changed since then, with the great growth in rural tourism and leisure, in counter-urbanisation, with the urban-rural shift in certain types of employment and with the diversification of farm household incomes (a majority of farm households these days have some non-farming income). However, policy and official structures have failed to reflect these changes and still largely view rural issues through an agricultural lens.

For example, both the mass media and government responded to the 2001 crisis largely as if it were simply an agricultural matter (as though we were back in 1967). A disease-control strategy that was 'ultra-precautionary' in order to protect the farming industry coincided with predominant news values (particularly the strong visual images of cows and sheep being shot and pyres of bloated carcasses) in determining the media's treatment of the crisis as an animal plague visited on the country.[5] Confronted with these grisly images and asked to stay away, the public obeyed, avoiding contact with farm animals, but also with market towns, village pubs and shops, country hotels and visitor attractions. The consequence was severe losses inflicted on the wider rural economy, which at least in the short term greatly outstripped those suffered by the farming sector.

The UK public and government have thus been rudely awakened to the diversity of the contemporary rural economy and agriculture's minor role within it. Leisure and tourism, manufacturing and services have replaced agriculture as the mainstays of local rural economies. Yet while demoting agricul-

15: Sheep grazing in Northumberland, the county in which the 2001 foot-and-mouth epidemic began and ended. (After foot-and-mouth)

16: Scene at the Northumbria food festival, 30–31 March 2002. Speciality food producers have recognised the need to collaborate to reach urban consumers. (After foot-and-mouth)

17: Livestock farming maintains most of the open landscapes of Britain. Without it, grasslands rapidly become rank and scrubby, losing much of their biodiversity. The cattle on the skyline are ready to set about restoring this grassland in a protected area on Sheppey, Kent. (The farmed landscape)

18: Properly managed, recreational lands such as the fairways, rough and semi-rough of golf courses like this one at Kingsdown, Kent, can maintain species-rich ecosystems. Downland like this needs appropriate mowing, with the cuttings removed, to maintain its flowery swards. (The farmed landscape)

19: Mineral workings, spoilheaps and other industrial sites can be restored to create new amenity areas and wildlife habitat. The spoil from the English half of the Channel Tunnel was deposited below Shakespeare Cliff, Dover, to create 90 acres (38 hectares) of new land named Samphire Hoe, seen here under construction.
(The farmed landscape)

20: On completion of the tunnel, the work site and spoil at Samphire Hoe was landscaped and revegetated to create a scenically spectacular recreation area rich in wildlife. (The farmed landscape)

21: A field invaded by young oaks, Colchester.
(Woodland and plantation)

22: All that is (or was) left of Mousehold Heath, Norwich, a famous
and beautiful heath that has been largely turned into an
uninteresting wood. (Woodland and plantation)

23: 'There was once a way through the woods.' A former lane between fields with walls around them, now lost in the woods that have spread over most of the state of Connecticut within the last 150 years. Is this the future for much of England? (Woodland and plantation)

24: Hayley Wood, Cambridgeshire, a 'bottomless' wood in which deer have eaten out the low vegetation. What is its future? (Woodland and plantation)

25: Ancient oak, Staverton Park, east Suffolk. Many of the values attached to trees depend on such wonderful, gnarled hollow pollards. (Woodland and plantation)

26: Signs of past depopulation at the abandoned township of
Lonbain, in the Applecross area of Wester Ross.
(The Scottish Highlands and Islands)

27: The crofting community of Elgol on the Isle of Skye, with the
Cuillin in the background. Peopled landscapes in no way detract
from the natural environment of the modern Highlands and Islands.
(The Scottish Highlands and Islands)

28: The designation of the Snowdonia National Park in 1951 recognised an area renowned for its beauty. Balancing the needs of visitors with those of the local communities is as delicate a task as the conservation of the natural assets. (Heritage landscapes in Wales)

29: Tredegar, at the head of the Sirhowy valley, developed in the late eighteenth century and expanded in the nineteenth century as coal mining became the mainstay of the economy. Like many of its neighbours, it now has to come to terms with the social and economic implications of post–industrialisation.
(Heritage landscapes in Wales)

30: The so-called 'Roman Steps', a medieval trackway that runs alongside Rhinog Fawr near Harlech, in an area where the landscape is strongly associated with myth and legend, fact and fiction. The trackway's traditional name is an example of the blurring of historical boundaries. (Heritage landscapes in Wales)

31: The marine colliery at Cwm, near Ebbw Vale, photographed in the 1970s before its closure in 1989 and the virtual obliteration of the south Wales coal mining industry. The once-familiar head frames can now be seen only in museum collections or at the last working Welsh deep mine, at Tower Colliery, near Hirwaun. (Heritage landscapes in Wales)

ture, the FMD crisis has also revealed starkly the continuing dependency of the countryside on farming. The rural economy may now be diverse and agriculture a minor component, yet it remains vulnerable to an agricultural crisis, and would still have been vulnerable even if the crisis had not been handled from such a single-minded perspective. This is because the predominant image of the countryside, and one that the crisis has tarnished, is a pastoral one, based on extensively grazed landscapes. That is what tourists and visitors appreciate. Agriculture's wider role in much of the countryside is thus mainly symbolic, aesthetic and ecological.

The particular sectoral incidence and geographical impact of the 2001 foot-and-mouth crisis highlighted the links between certain farming systems and the touristic countryside. Because FMD largely took hold in sheep, the heavily affected areas were those with extensive grazing systems and picturesque landscapes. What must be readily apparent now is that the public benefits of pastoral farming in such areas as Devon, the Lake District, the Yorkshire Dales and the North Pennines far overshadow the market value of its tradable products. An outbreak of FMD elsewhere – say in East Yorkshire or Lincolnshire (with their great concentrations of pigs) – would have had quite different resonances. After all, the swine fever outbreak in East Anglia in the year 2000, although much more devastating for the pig sector, did not have the ramifications for the wider rural economy that the FMD crisis has had.

More specific geographical dependencies and vulnerabilities have been revealed by the particular incidence of the FMD crisis. Firstly, since the mid-1980s, on-farm diversification has been promoted as a means of strengthening rural economies and boosting farm incomes. Such non-agricultural enterprises, though, were particularly affected by the quarantining of farms, which must raise doubts over the wisdom of this strategy. Secondly, the FMD crisis revealed the still narrow basis of the economy of some rural areas. The peripheral areas where the disease hit hardest – the north of England and the far south-west — are heavily dependent on primary industries and tourism, and consistently rank as the most deprived rural areas in England.[6] An important question facing future rural development policy is the extent to which a strategy of encouraging diversification from agriculture into tourism may risk simply shifting local employment from one vulnerable sector to another.

The FMD crisis, its conduct and its impact, raises profound questions about the relationship between agriculture and the rural economy, including how to secure sustainable agricultural livelihoods and how to promote more robust rural economies. The current system of agricultural and rural support was born out of the priorities and concerns of the 1940s and 1950s for food security and improved agricultural productivity. Recent changes to the CAP have been limited in their scope. The FMD crisis has focused attention on the need for further and more ambitious reform.

New structures and new approaches

New directions

Following the general election of June 2001, postponed at the height of the FMD crisis, the returned Labour government moved swiftly to set up new structures, including the establishment of the Department for the Environment, Food and Rural Affairs to supersede MAFF. Over the years the old Ministry had instinctively looked after the producer rather than the consumer, the environment or the rural economy. Foot-and-mouth proved to be its last stand. The new Department brings together agriculture, the food industry and fisheries (from the former MAFF) with environmental protection, rural development, the countryside and wildlife (from the former Department of the Environment, Transport and the Regions). In many respects it represents a welcome 'joining up' of policy: responsibilities that were at the margins of the former departments are now centre-stage, including the environment, sustainable development and rural policy.

It is noteworthy that this is the first time since the Board of Agriculture was established in 1889 that there has not been a ministry in central government with agriculture in its title. Necessarily, though, the new department retains strong links with producer groups, and some of the most pressing issues it faces are traditional MAFF responsibilities, including farm restructuring, the tightening of animal disease control, better food regulation and reform of the CAP. It is imperative, therefore, that the new department represents the broad public interest in the spectrum of issues it covers and that the interests of the environment, food consumers or the rural economy are not subordinated to those of primary producers. Achieving this crucially depends upon political leadership rather than administrative arrangements or the names of ministries.

Central to such leadership must be a new understanding of the role of agriculture in contemporary society. This role no longer rests on its contribution to GDP (less than 1 per cent) or to employment (just 1 per cent): in neither of these senses is agriculture still a significant economic sector. The far more crucial fact is that agriculture occupies and manages 80 per cent of the land: it thus constitutes the countryside and is the primary determinant of the state of the rural environment. Agriculture is also a vital and critical element in a food supply chain that needs to be much more integrated and responsive to the needs of consumers.

Agriculture, therefore, remains properly central to the new department, not as a stand-alone economic sector, but in its relationship to the rural economy and environment on the one hand, and to the food supply chain on the other. The task ahead is better to integrate agriculture into these broader functions. The industrialisation and globalisation of agriculture, though, have tended to detach farming from the rural economy and to marginalise or impoverish the rural environment. Certain contemporary consumer trends could help to reverse this process and foster a reintegration of agriculture into the countryside. Successive farming-and-food crises, for example, have heightened consumer concerns about the provenance of food and the way it has been produced. This is expressed through demands for quality assurance, traceability, organic production, welfare-friendly systems, and local and regional produce. All of these demands should serve to reconnect the final consumer with the primary producer by linking the qualities that consumers value in food to the methods that were used to produce it.

Opportunities are thereby opened up to develop markets and marketing systems that can deliver premium food products rather than cheap, undifferentiated commodities. Production systems would have to be environmentally friendly, and would be publicly supported in part because of this. Such environmentally friendly production itself provides valuable attributes to the stock or crop, but to earn income, these attributes have to be marketed. They therefore have to be presented as distinctive products, not as standard commodities.

One distinctive characteristic of any primary production is location, which cannot be duplicated or imitated by rivals in other parts of the country or the world. To turn this to competitive advantage requires the strengthening of the marketing chain to deliver an enhanced reputation of the origin – the craftsmanship, the socio-economic and environmental care – of the production and processing system. In turn, this will require innovation and imagination in product development and new promotion and marketing systems, with an emphasis on increasing consumer perceptions of the authenticity and quality of the product package.[7]

Creating an integrated rural development policy

Public support systems must be changed if the rural economy is to be strengthened. Chief among these support systems is the CAP. World trade talks and the commitment to enlarge the EU to Central and Eastern Europe themselves demand that the CAP be radically altered. The main desideratum is that anachronistic production subsidies to farmers be replaced with measures that assist the regeneration of rural economies overall. Under such an 'integrated rural development policy' the following arrangements would prevail:

- *First*, markets would largely determine the income that farmers receive from growing crops and raising livestock.
- *Second*, farmers would receive sufficient support for the environmental management functions of agriculture.
- *Third*, rural development would be given much greater promotion, to assist in the economic adjustment of rural areas and to help improve rural incomes and employment.

The priority must be to ensure that local rural economies become more robust, versatile and based on sufficiently diversified income sources. If they do, then this will allow farm businesses and households to adapt to changing economic circumstances. In this context it is important to dispel the increasingly outdated notion of the 'full-time farm business' – whereby the household is wholly dependent upon agricultural income – which has tended to serve as the dominant model for the agricultural sector. Multiple-income sources for farm households are now a widespread feature of British and European agriculture. Income sources brought in from outside the farm through some household members going 'out to work' (whether, say, in farm contracting, in the local service sector, or in a nearby town) are of much greater significance than non-agricultural income generated on the farm (through activities such as farm tourism or food processing, for example). The implication is that to diversify farm household incomes the most appropriate strategy is to stimulate diversification and economic growth in the rural economy (such a strategy has the additional benefit of assisting non-farmers too). This means that farm households are able to manage even when income from farming is being squeezed, provided the surrounding rural economy is buoyant and offers opportunities for alternative or additional income for farm household members. Thus the most pressing problems lie with those localities where the rural economy is too narrowly dependent upon agricultural production. It follows that the focus of intervention to promote rural development and employment should be the rural and regional economy and not the agricultural sector.

Rural diversification: the role of market towns

Although there seems to be some consensus that diversification is the correct strategy for rural areas, the concept can be defined in quite different ways, making agreement over what it means at a local level very difficult to achieve. For some, the basis of the definition is the word 'diversity'; hence the aim is broadly to ensure that the rural economy has a range of activities, that farm families have multiple-income sources and that school leavers have a choice of jobs. Such an approach follows from the view that the past overdependence of rural areas on a single sector narrowed the options and concentrated the risks too much. For others, the definition has more to do with transformation and the development of new and distinctive economic functions as rural areas redefine their comparative advantage in a changing world.

The crucial questions to be resolved are: What is our long-term goal for rural areas – are we aiming for a permanent state of hedging our bets? (Or are we in a temporary transition from one form of specialisation to another?) Is diversification something that can be left entirely to local decision-making and incentives, or should we be planning on a larger scale?

The approach to rural development of some other European countries has tended to be less piecemeal than ours, and has focused on local specialisation rather than heterogeneity. The

German approach, for example, concentrates on diversity *be-tween* areas of 'indigenous potential', encouraging local speciali-sation where a group of villages or a town has a natural advantage. For the Germans, diversification is therefore based on a systematic exploration of the possibilities for 'decentralised concentrations' of services and functions, aiming to distribute these between local areas. Such an approach makes the concept of diversification much more useful as a tool for rural develop-ment, and reduces the risk of local economies in rural areas becoming fragmented by small-scale diversification. It also recognises the fact that rural areas are on divergent socio-eco-nomic trajectories crucially shaped by their regional contexts, which means that opportunities for diversification are regionally and locally distinct. This suggests a more planned approach at the regional level than has traditionally been the norm in Britain. Without such an approach it is difficult to allocate public resources sensibly for such things as training programmes, infrastructural development, business advice and support, or re-gional economic promotion and marketing. Efforts to diversify the rural economy in a systematic fashion must, therefore, be carefully devised in the context of regional economic policy.

These efforts must also recognise some fundamental realities of contemporary rural life. One of these is that rural society is highly mobile. Some 83 per cent of rural households in England own a car, including 37 per cent that own two or more cars. A major consequence of this high level of car ownership has been the long decline of local services across villages in England. This, combined with the loss of local employment in the primary sector, means that the notion of self-contained villages – where people could live and work locally and rely on local services – is a thing of the past, however nostalgically appealing. More than two-thirds of employed rural residents in fact work in towns.

The battle is now on to secure the future of the small and medium-sized market towns. Traditionally, such towns have enjoyed a position at the heart of rural life. As the focal point for commercial and social activity and the administrative base for local and national government, they have served as centres of employment, retailing and service provision for centuries. They may also have been the hub for a particular economic sector, for example mining, textiles or manufacturing. However, some of these functions have gradually been undermined by the pace of social, industrial and agricultural change and many towns have witnessed the closure of their livestock markets and the loss of traditional employers. The increasing mobility and expectations of consumers have also put pressure on high-street retailers. At the same time, some services have been concentrated in cities and the larger towns, often increasing the difficulties faced by the smaller towns.

The picture is not all gloomy, of course. Many towns are adapting to the changing demands upon them, meeting the challenges that the modern world offers them and so continuing to thrive. They successfully provide the lifestyle benefits that people look for when living in smaller communities. Others, however, are not faring so well. It is important that investment resources be focused on these towns, therefore, to equip them with a range of modern services, and training and employment opportunities for their rural hinterlands.

Such a movement from a sectoral policy for agriculture towards a territorial policy for rural development in turn provides new possibilities for the reintegration of agriculture into the rural economy. In particular, the role of agriculture should be fully recognised in the creation of distinctive places and products in an increasingly homogenised world, and there should be greater emphasis on the green credentials of farming in product and place marketing, and on the contribution of agriculture to regional competitiveness, including its role in various supply chains (e.g. in the regional tourism, energy and food sectors). An agriculture that is thus more closely oriented towards local and regional markets and production networks undermines the established notion of agriculture as a separate national economic sector (with its own political institutions). In this way, the artificial divide between agriculture and rural development is broken down and farmers come to be seen and valued not only as food producers but also as 'rural business people' and 'local environmental managers'.

The farmed landscape:
the ecology and conservation
of diversity

Bryn Green

The British landscape is extraordinarily rich in wildlife, scenic beauty and recreational opportunity. It is possible, in a morning, to walk from mudflats and salt marshes through sand dunes, pine forest, oak woods, meadows, pasture and arable land, and out on to heaths and downs. I am thinking of Purbeck, but this experience could be repeated in many places. In some parts of Europe one might have to drive for hours to see such variety; in the United States, for days. The geology and geomorphology of Britain are deployed at a human scale and it is this that is the ultimate basis of its variety of habitats, diverse wildlife and the amenities of the countryside. But the geology alone has not generated this variety. Without human intervention most of the land up to the tree line on the mountains might still be covered in the forest that for most of the 15,000 years since the retreat of the ice caps of the last glaciation cloaked some three-quarters of Britain. Farming cleared the forest and has created and maintained an open countryside arguably far richer in wildlife and amenity than the primeval wildwood.

Where were the meadows of buttercups and daisies, the fields of poppies, the skylarks and lapwings before the forests were cleared? It is the gradual development of farming over

millennia that has permitted the largely spontaneous colonisation of cultural landscapes by indigenous species recruited from
naturally open habitats such as dunes, cliffs, wetlands and woodland glades grazed by wild animals. Apart from providing habitats for such species, the semi-natural meadows, heathland,
moorland and downland of Britain and northwest Europe
greatly extended the ranges of plants and animals. Heathlands,
for example, have affinities with Mediterranean shrublands;
some essentially Mediterranean species such as the Dorset heath
and the Dartford warbler reach their northwestern limits in
British heathlands. Downlands in many respects resemble the
great steppe grasslands of Central Europe and Asia, and again
have species, such as pasqueflowers and stone curlews, in
common with them. The moorlands of the northern hills share
species with the arctic tundra.

The familiarity to the European of cultural landscapes composed of aggregations of these semi-natural managed ecosystems should not obscure the fact that such landscapes are
virtually absent from those parts of the world where Western
human intervention is more recent. Even in seemingly comparable and superficially similar parts of eastern North America,
forest clearance and farming have resulted not in species-rich,
semi-natural ecosystems of native species but in species-poor
examples of meadow and pasture dominated by common European grasses and herbs. The famed Kentucky bluegrass is in fact
our common meadow grass *Poa pratensis*, while our familiar
plantains were known by the Native Americans as the 'white
man's footprint'. The farmland of New Zealand is likewise
almost entirely made up of European plants and animals; gorse
and broom are amongst the most familiar plants of the wayside.
Only in the forest and coastlands does New Zealand's extraordinary ancient indigenous flora survive. Just as in the United
States and much of temperate South America, the European
flora has ousted the natives from farmland.

In contrast, save for notable exceptions such as the grey
squirrel, very few species from these countries have established

themselves in Britain or Europe, and certainly nowhere near to the extent of establishing entire alien communities of plants and animals that dominate vast swathes of countryside. None of these countries has anything comparable to our semi-natural heaths and downs.

Today, however, these habitats are disappearing and European farmlands are becoming barren of wildlife, like farmlands in other parts of the world. Why should this be so? Why has the impact of Western European farming formerly been so relatively benign in Europe, yet so pervasive in completely altering landscapes when exported elsewhere? Why have British ecosystems been so relatively immune to the invasion of aliens from other parts of the world? Probably no other country has been exposed to such a great importation of exotic plants and animals into its parks and gardens. The majority have escaped at some time, yet remarkably few have naturalised. To answer these questions we must explore some of the ecological, historical and economic factors involved.

The ecology of biodiversity

A piece of downland turf a metre square might contain twenty to thirty species of flowering plant. The whole down might have twice that number, plus twenty-five associated species of butterfly. A modern, more productive pasture ideally contains only one species of plant – ryegrass – and no associated butterflies or, indeed, any other wildlife. To generate the richness of species on the downland three requirements must be met.

Firstly, a rich regional pool of species – the flora and fauna – must have evolved. Many parts of the world, particularly the tropics, with their great extent, stable climate and local geological and climatic variation, have very rich complements of species. The harsh conditions of polar areas support very few species. More temperate areas are generally intermediate in the richness of their flora and fauna.

Secondly, there must be an efficient dispersal of species. Even where a rich regional pool of species has evolved, some areas within the region may have received only a fraction of the regional species pool and have therefore very limited numbers of species. This applies particularly to small and remote islands, or island habitats such as lakes or mountains, which species find difficult to reach and in which they struggle to sustain viable populations. Britain has fewer species than similar parts of the European mainland and Ireland has even fewer than Britain. Ireland has, for example, no woodpeckers, or snakes. The latter has nothing to do with St Patrick, but is a result of Ireland having been cut off early from species recolonising from the south after the last glaciation. In this current inter-glacial period, which we optimistically call the post-glacial, the British Isles were isolated from the European mainland some 7,000 years ago. Many species, such as the Norway spruce and rhododendron, which reappeared in previous inter-glacials, failed to make it this time. Others are now lined up along the far side of the English Channel. Midwife toads, beech martens, cabbage thistles and even, surprisingly, birds and butterflies such as crested larks and poplar admirals, show what a barrier to colonisation twenty miles of sea can be.

The efficiency of dispersal of species, contingent on the location and size of areas to be colonised, applies at all scales. Studies have shown that in the lowland agricultural countryside of England a wood isolated by farmland needs to be at least a kilometre square (100 hectares) to have a good chance of containing all 48 species of lowland birds.[1] Smaller woods simply do not meet the requirements of the more demanding species. These requirements may be a particular facet of the habitat; nightjars, for example, need big clearings such as recently cut coppice. Or it may just be a question of population size. Small woods can sustain only small populations and small populations are notoriously susceptible to collapse under adverse conditions such as a hard winter. For both large oceanic islands and small island habitats, the number of species roughly halves as the size of the island reduces by a factor of ten. Thus a wood of 10 hectares might only hold 24 species of bird; a copse of 1 hectare only 12. Note that with the reduction in size it is generally the larger and the most demanding species that are lost. The smallest woods tend to contain only blackbirds, robins, wrens and other common species.

The third and final requirement for the establishment of diverse plant and animal communities is that each species should be able to corner its own particular supply of resources for survival – its niche. Many species compete for the same resources and under any particular set of environmental conditions one species will invariably have the edge over another. However small this competitive advantage, the most effective competitor will eventually out-compete and exclude the weaker competitor. Perhaps the most fundamental dictum in ecology is that 'complete competitors cannot co-exist'. This competitive exclusion tends to produce monocultures, dominated by the species best able to monopolise resources.

The best plant competitor on reasonably dry and fertile soils in northwest Europe is the beech tree. But if the soil is too infertile to supply enough of the nutrients needed to make a tree, or wind, fire or grazing animals kill the young trees, then the competitive advantage tilts to other species, usually dwarf shrubs like heathers, or to grasses. Disturbances or shortages of resources of this kind are therefore crucial in constraining the powers of the more effective competitors and allowing the weaker competitors to co-exist with the stronger in more diverse communities.

At Rothamsted Research Station in Hertfordshire a field experiment – the Park Grass Plots – has been used to study plant competition for over a hundred years. Various combinations and amounts of fertiliser have been added to a meadow to assess how grass production can be increased. The most effective treatments in increasing production have also been the most effective at reducing the diversity of the sward. Where added fertiliser has favoured the most vigorous and productive grasses, so they have competitively excluded most of the sixty or so herbs and flowers of the original meadow.

This is one of the main reasons why much of the colour has gone from the post-war countryside; why productive *Farmers Weekly* green ryegrass monocultures have supplanted tawny meadows sparkling with wild flowers. The objectives of the farmer, who generally wants high production, and the conservationist, who generally wants high diversity, seem to be fundamentally, diametrically and irreconcilably opposed. The conservationist interested in diversity often wants one blade of grass to grow where two grew before!

But it is not, of course, quite as simple as that. Both too much and too little disturbance, or too much and too little fertility, can also reduce diversity. Few species can survive under such regimes. The success of European species throughout the temperate world is probably due to the fact that they have co-evolved with domestic stock during millennia of farming and are better competitors than their native counterparts in North America and elsewhere, which have evolved under less intensive grazing regimes. When farm stock was introduced to these countries the plants of the native vegetation were outcompeted.[2] But if conditions should change to favour only the very strongest competitors, even most of the European species would find survival difficult. Thus, whilst farming, at least for much of its history, has been a force for creating and maintaining diversity, paradoxically, the subsidised, technological, productionist farming of the post-war period has had devastating impacts on wildlife and the countryside.

Human impacts

We know something of the nature of the post-glacial forest, its species composition and how it changed with time, from the pollen preserved in peat bogs. Its trees and other species varied as waves of colonising species arrived and were deployed onto suitable geology and soils. How continuous the tree cover was is less certain. Lack of grass pollen suggests there were then no extensive heaths and downs, or any other large open habitats, save at the coast or on mountains, where conditions are too severe for tree growth. Natural disturbances such as fire, flood, windstorms, avalanches and landslips can, however, all fell trees and create glades, which deer and other wild grazing animals can keep open. It has been proposed that in the past a park-like landscape, rather than thick woodland, may thus have been quite naturally maintained.[3] The great storm of October 1987 showed what a devastating impact natural disturbances can have on woodlands, but for most of the later post-glacial period at least, it seems that human disturbances have had the greatest impact.

The Middle Stone Age people who lived in Britain once the ice had gone probably did not live under the trees. Wolves, bears and, most fierce of all, aurochs, the now extinct ancestor of modern European cattle, made the forests dangerous places. People probably lived at their edges, especially around the shores of lakes and the sea. It was thus once thought that these Mesolithic peoples made little impact on forest clearance. More recent studies suggest that they began the process of pushing back the trees, probably through the use of fire to drive game. The tree line on the Pennines certainly seems to have been lowered like this and moorland initiated.[4] Some southern heathlands also seem to date back to Mesolithic clearances.

Extensive clearances, or at least the replacement of wild with domesticated grazing animals, did not take place until the introduction, some 5,000 years ago, of the most revolutionary human initiative of all – agriculture. It has been the most important force in moulding landscapes and determining the nature of the species that occupy them ever since. In Britain forest clearance was remarkably early and complete. It has been suggested that by the Domesday survey of 1086 only 15 per cent of forest cover remained.[5] By the beginning of the last century this had been reduced to 5 per cent. It is now about 10 per cent, the greater part of it forestry plantations. Given the level of environmental concern over the large-scale deforestation schemes currently taking place in developing countries, it might seem surprising that British landscapes still retain any wildlife or environmental features worthy of protection after the loss of so much original forest.

The early and comprehensive forest clearance in Britain, together with hunting, was probably responsible for the early loss of some of the larger mammals such as bears, aurochs (which survived in Polish woods until the sixteenth century) and beavers. Others, such as wolves and wild boar, almost survived to modern times in Britain and still do survive in similar landscapes on the continental mainland. After woodland clearance, wetland drainage has probably been the next most important cause of species loss and decline. The great flocks of wildfowl that darkened the sky disappeared with the reclamation of the East Anglian fens, the Somerset Levels and other great wetlands such as Romney Marsh. The Dutchman, Vermuyden, brought in by the Earl of Bedford in the seventeenth century to mastermind the drainage of much of the East Anglian fens, changed a way of life based on the exploitation of fish and fowl, peat and withy. Birds such as spoonbills were lost as breeding species, as the bogs, fens, shallow-water swamps and marshes gave way to first grasslands then arable. Small surviving island reserves such as Wicken and Woodwalton Fens give some indication of the nature, but not the scale, of what has been lost.

A third major impact on the countryside and its wildlife took place as a result of the Enclosure Acts, mainly between 1750 and 1850. Much of middle England, farmed under infield-outfield systems – with open arable crop fields in strips, riverside meadows providing hay for winter feed and heath or downland common grazings – was converted from peasant to capitalist farming by the complete redesign of the landscape into individual fields and farms. The enclosure of the outfield commons or 'wastes' incorporated into the new farm structures led to the loss of vast tracts of downland and heathland. With them went birds, such as the great bustards of East Anglia and Salisbury Plain, together with other species that require large stretches of open habitat. At least 1.42 million hectares and perhaps as much as 2.42 million hectares of commonland are estimated to have been reclaimed in England and Wales between 1780 and 1880.[6]

The beginnings of conservation

The loss of commons and of the open access to the countryside they provided was one of the main spurs to the development of the conservation movement in Britain. Through the formation of the Commons, Open Spaces and Footpaths Preservation Society in 1865, followed by the Royal Society for the Protection of Birds (RSPB) in 1889 and the National Trust for Places of Historic Interest and Natural Beauty in 1895, voluntary organisations began to press for legislation and acquire land to protect wildlife and the countryside. But, unlike in the United States, where government action began with the formation of the world's first National Park, Yellowstone, in 1872, it was not until 1949 that the National Parks Commission was established for England and Wales, to protect areas of natural beauty and recreational importance, and the Nature Conservancy was created to protect wildlife.

At that time National Parks, Areas of Outstanding Natural Beauty, National Nature Reserves and Sites of Special Scientific Interest (SSSIs) – all largely protected by the new powers to control development set out in the 1947 Town and Country Planning Act – were seen as the key measures required to protect the countryside from the perceived main threats of urban and industrial expansion. The thought that farming, not perceived as a threat at all, indeed regarded as being synonymous with the countryside, would come to constitute the major force in destroying countryside amenities was not even contemplated. As Lord Justice Scott put it in 1942:

farmers and foresters are unconsciously the nation's
landscape gardeners ... even were there no economic,
social or strategic reasons for the maintenance of
agriculture, the cheapest way, indeed the only way, of
preserving the countryside in anything like its traditional
aspect would be to farm it.[7]

The new powers and state support for the agricultural industry contained within the 1947 Agriculture Act were designed
to avoid a repetition of the agricultural depression of the 1930s
and the food shortages of the war. They proved, however, to
have unforeseen and disastrous environmental side effects. What
was true of traditional farming and of the depressed pre-war
landscape was not to prove true of the modern, thriving post-
war industry.

The nature of modern agricultural change

Concern about agricultural change and its social and environmental impacts is nothing new, especially at times of major
change such as the drainage of the Fens and during the Enclosures. John Clare, in *Remembrances*, conveys the sentiment:

Inclosure like a Buonaparte let not a thing remain
It levelled every bush and tree and levelled every hill
And hung the moles for traitors – though the brook is
    running still
It runs a naked brook, cold and chill

It is only in the last fifty years, however, that intensive, subsidised and technologically advanced farming, whilst greatly increasing production, has eliminated much of the biodiversity,
archaeology and even topography so widely cherished in the
British countryside. The idea that farming not only produced
food, but automatically delivered the kinds of landscape that

people want, was so deeply ingrained that this belief was not
seriously undermined even when, in the late 1950s, the use of
chlorinated hydrocarbon pesticides caused the collapse of many
species populations. Birds of prey, such as peregrine falcons, and
other species at the top of the food chains within which the
chemicals were concentrated, such as otters, were the hardest
hit. Realisation that all was not well with farming not only took
a growing catalogue of habitat, species and landscape losses, but
the revelations that: firstly, far from being short on food, British
and European agriculture was producing huge unwanted surpluses; and, secondly, far from enjoying a cheap food policy, the
consumer was paying two or three times the world market price
for many basic foodstuffs.

Post-war agriculture has been supremely successful in meeting its primary, paramount objective of increasing home production. Self-sufficiency in temperate foodstuffs has risen from a
pre-war 30 per cent to nearly 80 per cent today, but that success
has been bought at a high cost, both financial and environmental.[8]

Capital investment, the substitution of powerful machinery
for manual labour, highly effective pesticides and abundant,
cheap, synthetic fertilisers have all helped to transform the industry. The limiting factors of poor fertility, waterlogging, pests
and difficult terrain, which previously confined different kinds
of low-intensity production to particular kinds of land, are now
readily overcome with modern machinery and agrochemicals.
As a result simpler, specialist, more productive, large-scale enterprises have replaced much of the old crop, livestock, biological
and landscape diversity with homogeneous monocultures. Essentially, the agricultural changes have been of three main kinds:

   1 Specialisation and intensification. These have
      involved a greatly increased fertiliser use, chemical
      weed and pest control, a switch from mixed farms
      to livestock or arable farms, greater use of winter-
      sown crops, the elimination of stubbles, and

intensive livestock rearing, with a move from hay to silage.

2   The reclamation of marginal land. This has included the removal of hedges, ditches and headlands to increase field size, woodland clearance, drainage of wetlands, conversion of heathland, moorland and rough grazings to sown grassland, and ploughing of old grassland for arable.

3   The abandonment of marginal lands. This has mainly affected hill grazings no longer economic for livestock rearing, which become rank grass and scrub, but also coppice woodlands, hedges, ponds, paths, and other farm features no longer part of the agricultural system.

Environmentalists were initially slow to begin to monitor and assess these impacts, but they are now well documented and accepted, though there is perhaps still some scepticism in parts of the agricultural industry.[9] The first comprehensive surveys undertaken for Great Britain showed that, between 1949 and 1984, 30–80 per cent of habitats such as upland moors, lowland heaths and rivers, ancient woodland and limestone grassland were lost or significantly damaged[10] and, for England and Wales, that between 1947 and 1985 overall hedgerow length fell by 22 per cent and semi-natural vegetation cover by 25 per cent.[11] More recent, more systematic surveys show that, despite attempts to address these changes by encouraging more environmentally friendly farming, the losses continue. Between 1984 and 1990 hedgerow length again fell by 23 per cent and over a similar period farmland bird populations have suffered substantial declines (see Table 5).

The loss, fragmentation and small size of surviving habitats have been a significant factor in species decline, but agrochemicals continue to have indirect adverse effects on species populations. Pesticides, but also herbicides, by eliminating the weeds on which insects feed, may not affect bird populations directly but

Table 5 **Percentage of population decline in farmland
         birds, 1970–95**

Tree sparrow	89
Grey partridge	82
Corn bunting	80
Turtle dove	77
Bullfinch	76
Song thrush	73
Spotted flycatcher	73
Lapwing	62
Skylark	58
Linnet	52

Source: DETR (1998), Campbell and Cooke (1997)

do remove their insect food. This has been shown to be the main cause in the decline of partridge populations.[12]

Fertiliser nitrogen, in an even more insidious manner, is perhaps the greatest threat of all to biodiversity. Much leaking of nitrogen takes place from farmland into surface- and ground-waters and, through dust and denitrification, into the atmosphere. The consequent eutrophication (nutrient enrichment) of both water and land favours the more vigorous species, which out-compete others and so reduce diversity. Blanket weeds now choke waterways, while nettles, thistles, cleavers and cow parsley dominate hedgerows and bracken invades hill pastures once full of wild flowers. In some parts of Britain up to 50 kilograms of nitrogen per hectare per year now fall in rain, perhaps ten times previous levels. Not all this is attributable to farming – industry, car exhausts and many other kinds of pollution supply NOX (nitrogen oxides) as well as SOX (sulphur oxides) – but farming is nevertheless a major contributor. Fertiliser chemicals are essential for farm production. Though free of pesticides, even organic farming depends on natural fertilisers and these can be just as polluting as synthetic nitrogen.

Some species in habitats away from farmland have increased their populations against the general trend. Birds of prey, such as peregrine falcons and sparrow-hawks, have recovered well following bans on chlorinated hydrocarbon pesticides. Many wetland birds, such as avocets, the emblem of the RSPB, have recolonised with the provision of new wetland reserves, reservoirs and the wet pits of the gravel industry. Neither has farming been the only agency to cause environmental problems. Urban and industrial expansion, mineral extraction and conifer afforestation have all led to species and habitat loss. All these land uses have, however, also had some beneficial environmental effects. In contrast, the impact of modern farming has been almost entirely negative and the fact that it occupies some 80 per cent of the land surface makes it by far the most damaging environmental force.

Environmentally friendly farming

A desire to control agricultural overproduction and the enormous cost of farm subsidies under the Common Agricultural Policy (CAP) of the European Union led to reforms in 1985 and 1992 that, for the first time, switched some support funds from production to encourage environmentally friendly farming. These agri-environment schemes were developed from measures pioneered in Britain. The first effective payment of farmers to manage the environment was made possible by the 1981 Wildlife and Countryside Act in the form of 'profits forgone' compensation, a means of reimbursing farmers for loss of production and consequent profit when they were prevented from undertaking environmentally damaging developments in SSSIs and National Parks. Subsequently, an experimental scheme at Halvergate Marshes in Norfolk, undertaken by the former Countryside Commission, moved the emphasis from very costly 'profits forgone' compensating payments to positive payment for continued marshland grazing. This principle has underlain the plethora of environmental schemes that have subsequently been developed under the CAP.

The two main agri-environment measures in Britain are Environmentally Sensitive Areas (ESAs) and the Countryside Stewardship Scheme (CSS). ESAs are geographically delimited parts of the country important for their scenery, wildlife and historical features, such as the South Downs, Pennine Dales and Somerset Moors and Levels, which are 'threatened by changes in farming practices'. In them farmers can enter into ten-year agreements to improve the environment by, for example, maintaining hedges, flower meadows and wetlands through a variety of activities, ranging from reduced use of agrochemicals to the raising of water levels. Countryside Stewardship involves similar practices, but applies everywhere and has the crucial difference that DEFRA does not have to accept all its applications. This discretionary element means that money can be targeted to prioritise features or places that are particularly threatened.

All these schemes are voluntary and their take-up depends upon the circumstances and inclinations of individual farmers. Some schemes, notably Countryside Stewardship, have been oversubscribed. Others, including the more demanding tiers of ESA measures, such as those for the return of arable to grass, have not been taken up so enthusiastically. In Britain approximately £100 million of agricultural support has been switched to these agri-environmental schemes over each of the last few years. Whilst there is no doubt that this is a significant injection into environmental management, it is only a small fraction of the total agricultural support of £3 billion per annum. The consensus of environmental opinion is that these measures have simply slowed the rate of environmental losses and that much more farm support needs to be redirected to environmental management if these losses are to be stopped and environmental improvement is to be attained.[13]

The transfer of more money to environmental management is now seen by many, including farmers, as perhaps the only way in which some state subsidy of the industry can continue, given international trade liberalisation, overproduction and the consequent decline in farm markets and product prices that has led to the recent calamitous fall in the profitability of the industry. In particular, such a transfer is seen as perhaps the only way in which many small, marginal farmers, mostly involved in livestock husbandry in mountain areas, can be guaranteed future livelihoods. But this has still to be reflected in any significant way in proposals to reform the CAP. Whilst some European member states have reallocated significant amounts of production subsidy to environmental payments (in Austria it is some 30 per cent, with 70 per cent of farmland under agri-environment agreements), most have not done so. Presently only some 4 per cent of agricultural subsidies are directed to environmental management schemes in the EU as a whole, where 17 per cent of land is under agri-environment agreements. In Britain these figures are respectively 3 per cent and 6 per cent.

Farmers as countryside managers

Only farmers have the equipment, stock, knowledge of the land and, most crucially, its tenure, to manage the countryside. Some farmers have always farmed in environmentally friendly ways for their own satisfaction. Organisations of these farmers, such as the Farming and Wildlife Advisory Group, have done much to promote environmentally friendly practice. If all farms were as rich in wildlife as those of these farmers there would be few concerns over agricultural impacts on the environment. Unfortunately, most farmers seem to have kept their enthusiasm for environmentally friendly farming firmly under control. Nonetheless, as agri-environment schemes have undoubtedly slowed, and in some cases reversed, the attrition of the environment by intensive agriculture, many farmers and environmentalists are confident that more money for agri-environment schemes would solve the problem entirely. But there are several reasons that suggest their confidence may be misplaced.

Firstly, it seems unlikely that the large-scale continuation of agricultural subsidies, even with environmental rather than production objectives, will be acceptable under international trade agreements. Some switching of subsidy funds may be acceptable to our World Trade Organisation partners but probably not of a large proportion of the entire billions of ECUs of production support. Countries such as Australia and New Zealand have long removed all their subsidies to farming, yet still compete successfully in world markets. The removal of farm subsidies and the exposure of New Zealand farmers to world prices in 1984 did not lead to the predicted rural collapse; rather it reduced land prices, making for easier entry of young farmers into the industry, diversified cropping patterns, reduced sheep numbers and led to more forestry on marginal farmland, greater biodiversity and lower food prices.[14]

Secondly, over half the farmers in the EU are over 55 years old and nearly half have no successor. Marginal farming, particularly uneconomic livestock production in the hills, though so important in maintaining open landscapes and seemingly such an attractive way of life when viewed from the perspective of the summer visitor, is in fact a very hard way of making a living. The recent crisis in the industry, compounded by BSE and foot-and-mouth disease, has hit these farmers the hardest, with the result that young people are moving out of such farming everywhere into urban and other easier lifestyles. In Britain an NFU survey has suggested that 85 per cent of hill farmers have no successor. Some back-migration, especially of hobby farmers and other part-timers, is to some extent compensating for this in certain areas, but without an adequate workforce such farmlands must either pass to forestry, to other uses, or be abandoned.[15]

This has already happened to a surprising extent in Europe. Between 1965 and 1983 the utilised agricultural area in the EU fell by 8 per cent, whilst forestry land increased by 15 per cent.[16] In the east-coast states of the USA vast tracts of farmland have been abandoned to forest. Tree cover in New York State, for example, has increased from 15 per cent to 65 per cent in the last fifty years.

Thirdly, there are considerable administrative problems and costs involved in devising, monitoring and implementing agri-environmental schemes that are equitable between different kinds of farmer, do not interfere with market competition, avoid corruption and reward farmers pro rata for work undertaken or objectives achieved.[17] Performance indicators in ESAs, for example, have relied much more on take-up acreages than on field measurement of environmental change.[18]

Fourthly, all the agri-environmental management agreements in place between the government and farmers are for limited periods, generally of ten years. What then happens if, on termination, a farmer decides he must plough the meadow and the flowers he was previously paid to protect? This has occurred on the expiry of some agreements and it raises the question of whether some form of planning control is necessary to protect this enormous public investment. Some have pointed out that it would have been much cheaper for the state to simply buy the land.[19]

Finally, even were there sufficient labour and money, it would surely not be desirable for the entire countryside to be made into a museum of ersatz landscapes, with a hundred years' worth of technological progress in agriculture abandoned and traditional techniques frozen in an unchanging time warp. This might conceivably be possible if all our competitors in world markets were to accept the same limitations, but that is highly unlikely because, as discussed earlier, they do not enjoy the same benefits of cultural landscapes, varied and shaped by a multiplicity of human uses, as we do in Europe. Strangely, however, such a scenario does not appear to be all that far removed from the objectives of some conservation organisations still seemingly locked into a vision of the British countryside of the 1930s.

Whilst it may be possible to maintain substantial, representative landscapes, or at least their cherished features, by more environmentally friendly farming, this will probably require more than just agri-environmental schemes. Such incentives need to be backed up by environmental regulation of conventional agriculture. This might be achieved by systems of cross-compliance, whereby subsidies, so long as they survive, are made conditional on environmental care. More planning control probably cannot be avoided in an industry that has an unusual history of freedom from control. New trends in food consumption may also help to promote the continuation of traditional farmed landscapes.

Green consumerism

Farmers have blamed the low prices they receive from supermarkets, compared with those on the shelves, as a key factor in the decline in farm profitability. State intervention in agriculture, such as the guaranteed prices and markets of the CAP, with all production, including surpluses, being brought into intervention, arguably atrophies the marketing skills of agricultural producers.[20] There are, however, encouraging signs of improved market orientation through the production of specialised niche products of local provenance and their sale through farm shops and farmers' markets.

Growing consumer demand for higher-quality, traditional foods produced in less intensive, traditional ways may to some extent help maintain farmers as wildlife and landscape managers. This addition of value to basic production has long been a feature in France, where consumers are prepared to spend a higher proportion of their disposable income on food. Such products demand and command premium prices, generally being produced by more labour-intensive and environmentally friendly methods, and aimed at the wealthier market sectors. The sheep whose milk produces costly Roquefort cheese, for example, help maintain the wonderful Causse grasslands of the Cévennes National Park. In Britain the Countryside Agency is promoting this approach through its 'Eat the View' campaign.

The purchasing power of supermarkets, despite its potential to depress farm prices, provides a most powerful mechanism for the expression of consumer preferences. If the link between low-output, high-value food products and farming practice can be made in the minds of consumers and the products prove capable of seizing a sufficiently large sector of the market, then this should improve the viability of traditional farming systems. Much must depend on how competitively these niche products are priced against more conventionally produced foodstuffs and on how willing consumers will be to pay whatever premium is necessary. In the UK the growth in demand of the organic food sector, particularly through supermarkets, now outpaces home organic production. Public concern about the development of genetically modified organisms in agriculture, whatever its true nutritional or environmental implications, seems to be an important factor in encouraging consumer demand for organic products. Yet, properly exploited, genetic engineering also has the potential to be environmentally benign by, for example, reducing agrochemical use.

The coming together of food production and marketing, where farmland is controlled, indeed often owned and managed, by large retailers, may offer the best prospect for low-input, low-output, high-value foodstuffs produced in environmentally friendly ways. Some supermarkets already operate such quality-assurance schemes. Tesco's Nature's Choice range involves a farm wildlife and landscape survey and Sainsbury's require some of their major fruit and vegetable suppliers to prepare Farm Biodiversity Action Plans.[21] Through very large-scale operations, including those of large-scale farms, economies of scale may help lower prices to levels that will greatly widen the environmental market sector. This may not help small farmers, but it may increase the extent of land that is appropriately farmed. The future of small farmers probably lies in part-time and hobby farming. Contrary to accepted wisdom, the loss of traditional small farmers may not be as damaging to the environment as commonly imagined. Small farmers often seem to farm in environmentally friendly ways largely by default, since they lack the resources to farm more intensively.[22]

What future for farming?

> Sufficient numbers of farmers must be kept on the land.
> There is no other way to preserve the natural
> environment, traditional landscapes and a model of
> agriculture based on the family farm as favoured by
> society generally.[23]

Whether such sentiments, echoing those of the Scott Report
made nearly half a century earlier, are feasible today is another
matter. Since the Mansholt report of 1969, studies have suggested consistently that potential surplus food production in
Europe is enormous.[24] A Dutch government report argues that
as much as 80 per cent of European farmland could come out of
production by 2015, and yet self-sufficiency in food be maintained, provided the remainder were to be intensively
managed.[25] The accession of Eastern European countries to the
EU will greatly increase food production capacity. On the other
hand, with the liberalisation of trade, there could be the potential for Europe to increase its food exports.

Whilst overall world food security is probably not nearly as
threatened by population growth as pessimists foresee, the
feeding of some countries, especially those in Africa, would
seem likely to depend upon Europe joining North America as a
major exporter of foodstuffs.[26] If world food supplies do come
under pressure, most of the demand will be for grains, which
produce more food energy per unit area and unit cost than
animal products. Although, therefore, the prospect of vast areas
in Europe being released from farming on the North American
pattern is unlikely, the existing polarisation of agriculture, with
more intensive crop production on the better land while marginal grazing lands are managed less intensively, afforested or
abandoned, looks set to continue.

Modern agricultural practices cannot automatically deliver
the landscapes rich in wildlife, scenic beauty and recreational
opportunity provided by traditional agriculture – they make the

land too fertile for all but crops and vigorous weeds. Ecological principles dictate that to create and maintain species diversity either some production has to be sacrificed, or productive and protective areas zoned separately at field, farm or local level. Much of the wealth of wildlife and the variety of the old farmed countryside of Britain was an accidental by-product of an inefficient, unproductive agriculture zoned at the field scale. Supplies of fertiliser were then at a premium and the outfield moors, heaths and downs were systematically robbed of their fertility to maintain the cropland by folding sheep. If ways can be found to make such lower-output farming with wild outfields, headlands and hedges profitable through the production of high-value products, we may be able to maintain some of the best of these old landscapes, or at least the best of their cherished features. Small-scale organic farming, being of necessity mixed crop and livestock, achieves some of these objectives, with the important additional environmental benefits gained through the restricted use of agrochemicals. Larger-scale organic enterprises can, however, be just as polluting as conventional farms in terms of their eutrophication of the environment with nitrogen.

On the better land, the new technologies and practices of 'precision farming', such as integrated pest management, genetically engineered crops and satellite mapping of patterns of field production levels, could also all reduce agrochemical inputs whilst maintaining yields. Whether the risks of 'genetic pollution' of wildlife outweigh the potential benefits of genetically engineered crops remains worryingly unclear, and consequently most environmentalists advocate prudence in their use. Non-food crops for fibre, pharmaceuticals and biomass energy will likely replace food crops in surplus, as rape and linseed are doing already.

In some places flood- and erosion-control cropping could become recognised forms of land use, as they have long been in the United States. Regulation through zoning of farm enterprises and practices to protect aquifers from nitrate pollution is already in place in designated Nitrate Sensitive Areas. Arguably something similar should protect alluvial flood plains from field drainage and coastlands from incursions of the sea. The disastrous floods of the winter of 2000/01 were probably attributable in part to overzealous channelling of water from farmland that would have previously lain flooded for much of the winter, recharging aquifers depleted by overexploitation. A pilot scheme is already under way in Essex to reopen coastal farmland to the sea to accumulate sediment and assist coast protection.

The farmed landscapes of the future are thus likely to be more multi-functional than those of the immediate past. How the new demands on farmland additional to food production will mould the composite whole remains to be resolved. Experience suggests that change will always be resisted and regretted, but the new eventually accepted.

New lives, new landscapes

> A negative policy of not disturbing the old cannot ...
> for long succeed. We must disturb it to survive – on a
> vast scale and everywhere ... In the period since our
> landscape was created the changes have been more
> sweeping than in thousands of years before, yet the
> translation of social change into changed environment
> has still barely started ... [27]

Future demands on land are likely to result not only in farmland becoming more multi-functional, but also in less land being farmed and thus passing to other uses. Big changes in European rural land use have already taken place in the second part of the twentieth century, with tourism now being a much more important base to the rural economy than agriculture in many European mountain and coastal areas. New leisure landscapes, such as nature reserves, national parks, golf courses, equestrian centres, marinas, leisure parks and community forests, are the twentieth-century equivalents of hunting estates and lodges, but open up the countryside to a much wider selection of people, particularly those from urban areas.

Agricultural overcapacity, coupled with a demand for 700 new golf courses in the UK, as perceived by the Royal and Ancient Golf Club of St Andrews, led to the construction of 400 new courses between 1992 and 1996, many on retired farmland. These golf courses are at present generally far richer in wildlife than farmland, though the introduction of agronomic practice in their management does not augur well for their wildlife – if treated with herbicides and fertilisers like farmland, they too will lose biodiversity.[28] New kinds of land use create new kinds of landscapes and those demanded and generated by an increasingly urban society are likely to be very different from those of the old rural society, as growing conflicts over access to private land, animal welfare and transport networks are already beginning to reveal.

The old wildlife- and amenity-rich countryside of flowery fields, coppiced forest land, wooded pastures, downlands, heathlands and moorlands was the incidental product of a rural industrial society. Some of the new urban-centred landscapes, such as golf courses, reservoirs and the wet pits of the mineral extraction industries, seem to be almost as congenial for wildlife or leisure as were earlier rural industrial landscapes – they are certainly richer than most modern croplands. As the new agricultural landscapes do not automatically generate biodiversity and amenity, they offer stern challenges, together with the slag heaps, quarries and slurry beds of the extractive industries and areas of urban dereliction, to those concerned with realising some of their potential for amenity use.

Countryside managers are rapidly improving techniques for habitat restoration and translocation to meet these challenges. Whilst it should not be supposed that ecosystems that have taken centuries to develop could be instantly replaced, habitats of considerable amenity value can be created quite quickly. The spoil from the Channel Tunnel, for example, has created a spectacular 38-hectare park beneath the Dover cliffs that was vegetated with native species and soon colonised by thousands of the rare early spider orchid.

This new public open space, Samphire Hoe, is managed for Eurotunnel by the White Cliffs Countryside Management Project. As with other similar projects throughout the country, this management body is a consortium of local authority, government agency, business and voluntary organisation partners, working together to improve the local environment and encourage economic regeneration by involving local people and raising environmental appreciation, enjoyment, concern and care. These initiatives developed from the 'Groundwork' projects originally set up by the old Countryside Commission, which were designed to achieve the same objectives in depressed urban areas.

Such countryside management schemes have been a great success in motivating and harnessing the energies of volunteers in both small- and large-scale undertakings, from tree planting to transforming urban wastelands into parks and species-rich habitats, whilst performing an important educational and economic service for local communities. The new national forest around Leicester and the ten community forests around other major cities will convert thousands of hectares of urban fringe land to complexes of parks, farms and open woodland. In total, public open spaces, parks and gardens in urban areas constitute a major environmental resource and schemes like the above are ensuring such resources are now much more appropriately managed. As a result, towns and suburbs are arguably already much richer in wildlife than most of the farmed countryside and perhaps destined to become even more so.

Many of the landscapes of the old countryside evolved slowly through the sum of the individual management decisions of numerous farmers and landowners. Others were redesigned or created anew and were therefore imposed more rapidly. In much of Britain the enclosures transformed an open field landscape into a planned rectangular layout of fields and farms. In the Netherlands sea defences, drainage and land

reclamation converted the estuaries of three great rivers into
the polder farmland of a nation state. In the United States in
the 1930s depressed marginal farmers in the hills of West Vir-
ginia were resettled on better land and their pastures and orchards allowed to revert back to the wild. A new tourist
access road, the Skyline Drive, was constructed. Now the
beautiful second-growth forests of Shenandoah National Park
attract more visitors than almost all other national parks in the
United States. It has been suggested that a similar development might revitalise the economy of rural Wales.[29] There is
enormous potential in Europe today for similar landscape
transformations. The Dutch, with most of their country reclaimed from the sea, are, unsurprisingly, the most advanced in
making and conceiving imaginative new landscapes specially
designed to accommodate new lifestyles.[30] Some innovative
approaches are being developed to explore alternative ways in
which familiar landscapes might develop and to assess people's
preferences for them.[31]

Given the current crisis in livestock farming in the UK,
particularly in the hills, where it has survived since the Second
World War only by courtesy of subsidies, the maintenance of
open moorlands, heathlands and downlands is likely to prove
the greatest challenge. Despite decades of careful conservation
management, scrub and tree encroachment is already advanced
on most surviving heath and downland. Scrubbing over might
be slower on the northern hills, but in the United States what
was once similar open pastureland is now forest, recolonised by
bears, beavers, wolves and mountain lions. Only the dry stone
walls running through the woods betray their former existence
as farmland, 50 to 100 years ago. But it is not much fun leaving
the car park and climbing a 1,000-metre mountain just to see
no further than the next tree – we might not want the Lake
District to be entirely covered in forest like the similar terrain
of New Hampshire. Yet our bald hills would surely benefit
from some more trees. Given that New York State has 7 million
hectares of forest in an area smaller than England, a doubling of

the current British forest area of 2.5 million hectares can be contemplated with equanimity.

Should lower land prices stem from the farming depression, it may be possible to acquire large tracts of marginal farmland and maintain them as open grass or heathland, with feral or semi-feral stock. This has been successfully achieved by the Dutch in their great Oostvarderplassen reserve, which is grazed by feral herds of wild cattle and deer.[32] It is also essentially the basis of the management of the New Forest, one of the finest natural landscapes remaining in Europe. Controlled burning can also be an effective management tool in such areas and, despite the reluctance of amenity land managers to burn, the traditional swathing of moorland needs to be maintained and more widely deployed. The reintroduction of species such as beavers, wolves and wild boar, which is now contentious, might be more acceptable in such large wild areas. Wild boar have already accidentally recolonised Kent and Sussex woodlands, after escapes when the great storm of 1987 brought down boar farm fences. Despite some alarm, they seem to survive without causing anyone much trouble, as they have always done across the Channel. Other species, such as sea eagles and kites, a century ago reviled and exterminated by stock-keepers, have already been successfully reintroduced and a tremendous window of opportunity now exists to restore many more species and create new wilderness areas in Britain.[33]

Such polarisation of land uses, and the transfer of land from productive farming, might require substantial mental gymnastics for many of us. It is a vision far more radical than the worthy, but mostly unimaginative, proposals of the recent white paper 'Our Countryside: The Future'. The received wisdom is that if farming were to abandon some of the land, then awful scrub and dereliction would be the inevitable result. The outcome is almost always discussed in pejorative terms. But it need not be so. Admittedly, a more multi-purpose countryside may presuppose a little less exercise of landowning powers and more public rights, which runs counter to recent trends. And yet, almost uniquely to the UK, there is a considerable post-war history of reshaping private property rights, starting with the planning control of the right to develop – a concept those from the United States find remarkable.

Although modest in scope, the new right of access to private land planned in the Countryside and Rights of Way Act is nevertheless the one really radical proposal in the white paper, and one that represents a swing back to greater public use. With some anticipated repatriation of CAP powers, the means of reforming agriculture and rural land use may not be out of reach. We should not simply be looking back to landscapes of the past, but taking the opportunity to conceive, design, create and maintain new landscapes fit for the social, economic and environmental needs of the twenty-first century.

Woodland and plantation

Oliver Rackham

Anyone engaged in making predictions for the next twenty years should look back twenty years to see the outcome of the predictions then made. In 1981 I was working towards *The History of the Countryside*, published in 1986. In the conservation chapter of that book I lamented the unprecedented changes, nearly all of them destructive, in the landscape since 1945, and the feeble efforts of conservationists to oppose them. Many ancient woods had been grubbed out to grow cereals, and even more had been felled and poisoned, and replaced by plantations, usually of conifers (a process euphemistically called 'restocking').

In the 1950s and 1960s there was an air of inevitability about these changes. The arguments for them were usually presented in economic terms, and conservationists, seldom able to distinguish good from bad economics, dared not challenge them. Even National Trust ownership did not then prevent woods from being replanted. The most that could be hoped for was to withdraw small areas to be used as nature reserves and to plead for cosmetic concessions, such as leaving the edges of woods unplanted or making the edges of new plantations sinuous rather than straight, to mask the destruction of the countryside at large.

However, by 1981 the end of these dark years was in sight. Much of the destruction was turning out not to have been worthwhile, even for its original purposes, and an appreciation of historic landscapes was developing. There was an improvement in the prospects for ancient woods, not only because there was less money to spend on destroying them, but because land-owners were beginning once again to understand and take pride in their woods as they were.

When I wrote *The Illustrated History of the Countryside* in 1993, the conservation chapter was the one needing most drastic revision. After thirty years of simple destruction, there had been a dramatic change in favour of conservation. Destroying wood-land now needed a stronger justification than that of merely growing a little more wheat or pulpwood. 'Remaking the land-scape', in the sense of the 1960s and 1970s, had become a thing of the past. The landscape had reverted to something like its normal dynamics of human adaptation and natural alteration. Ancient woods were listed and valued in their own right, not simply as vacant land that ought to be doing something useful. Even small-scale destruction of woodland, for example by road works, met with vigorous and often successful public resistance. A movement had begun to rehabilitate replanted woods.

Conservation has gone from strength to strength, and it has gone on beyond conservation itself to embrace attempts at restoration of lost features – to the extent (on occasion) of re-planting hedges in the exact pattern that existed before 1950. Even the recovery of Whittlesey Mere, destroyed in 1851, is not beyond the bounds of possibility. Very little ancient woodland has been lost in the last twenty years, and this has been out-weighed by ancient woods, once thought lost to replanting, that have been recovered: an example is the northern two-thirds of Chalkney Wood, Essex.

Farming in retreat

There is still pressure on land: ancient woods are threatened from time to time, and doubtless always will be. But in general the pressure has diminished. This is not just a phase in a boom-and-bust cycle: the fundamental cause of the reduced pressure is plant and animal breeding leading to greater productivity and therefore the need to farm less land. For many years farmers have been protected from the consequences of this breeding by European Union subsidies and by the poor state of what would otherwise be the competing agriculture of Eastern Europe, but neither of these is likely to go on for ever. This pattern may not continue, because plant and animal breeding has taken a direction ('genetic modification') that runs into public opposition and prohibition. But even if the breeders never extricate themselves from this position, there will still be the question of what to do with the land released by the higher-yielding varieties of the 1950s to 1970s.

As yet the decline of farming has not gone so far as to cause a significant abandonment of land in Britain, though matters got suddenly worse, especially in north and west England, with the foot-and-mouth disease of 2001. Mountain farmers, having given up arable farming in the nineteenth and cattle in the twentieth century, and having received poor prices for their sheep in the 1990s, are unlikely to persevere for ever in the face of an artificial disaster that may recur at any moment.

If, as seems likely, large areas cease to be farmland, what will become of them? Britain in the last few decades has shown an alarming tendency to eat up land with low-density development on the American model: Milton Keynes is a notorious example. But most of this happens through developers outbidding farmers in areas where farming is still viable; it will not impinge on abandoned farmland in more remote areas.

New natural woodland

Most fields, downs, and moorland in Britain are non-woodland because people and livestock put yearly effort into preventing trees from growing there. If this effort is withdrawn, these areas will turn into woodland. This should cause no surprise: anyone can see it happening on railway earthworks all over the country.

Succession – the processes whereby abandoned farmland and industrial land turns into woodland – was the fashionable theme of ecological research in the 1920s and 1930s.

This has happened before. Much of the woodland arising in the early twentieth century still exists, notably in the plotlands (remains of attempts at low-density housing) of south Essex. The historic heaths of Surrey and other southeastern and eastern counties are also now mostly woodland. Smaller areas of wood-land, overlying the ridge-and-furrow of ancient cultivation, are reminders of previous agricultural depressions going back to the Black Death or even earlier. Outside Britain, fields and terraces have turned into woodland on a very grand scale in eastern North America, the north Mediterranean, and even in the mountains of Japan.

Should conservationists oppose this change? The public is encouraged to believe that woodland is good – any kind of woodland except 'scrub', which is young woodland and is bad. Whether the public would still believe this if woodland were endless remains to be seen. However, woods are not all the same, nor do they all replace the same type of land. Where arable land or 'improved' grassland turns into woodland the result is often a gain for conservation. But all too often the opposite happens: heath, old grassland, or fen of the highest conservation value turn into non-distinctive woodland.

Bodies like the National Trust have resisted the change. They and their tenants have made heroic efforts to keep farms going in order to maintain a semblance of the 'traditional' cultural landscape of Britain, albeit in the simplified form that keeps only sheep. In the United States the Amish sect, required by their religion to live a modest lifestyle as farmers, are holding back the return of the forest from the landscapes of Pennsyl-vania, but it is hard to imagine that mere economics, unassisted by spiritual forces, can resist the pressures of abandonment except on a very small scale. In other parts of the eastern United States endless woods, full of field-walls and remains of hedges, cover the landscape, except where an occasional farm has been kept going as a historic feature, a demonstration of an otherwise vanished way of life.

Plantations

From 1920 until 1975 the Forestry Commission and its follow-ers were obsessed with conifer plantations, both on former farmland, heath, and moorland, and as a replacement for ancient woodland. This had its origin in various practical, political, and ideological motives, but in the latter part of the period the growing of conifer plantations was presented as an economic necessity. Elaborate predictions were made of the financial return from growing Sitka Spruce or Corsican Pine in relation to the money invested in destroying the previous vegetation and establishing the trees. As an unbeliever, I used to pick holes in the arithmetic, especially the big hole left by the fact that the discount rate, needed to equate money spent now with income to be earned forty years on, was plucked from the air.

Now that the forty years are up the prediction has failed for wider reasons. It was based on three gambles: that the planted trees would grow; that the cost of maintaining them would remain constant; and that there would be a market for the trees once they had grown. In reality the planted trees have usually grown (although they may not have prospered), provided they were on non-woodland sites, but have very often failed where they were the successors to a previous wood; men to weed and thin them (if they can be found) have demanded higher wages; and ordinary coniferous timber has had to be sold against foreign competition. Not surprisingly the traditional suppliers in the Baltic – well organised and with 700 years' experience in shipping timber to Britain – can provide timber more cheaply than if it is felled here. Trees can still be sold, but they do not sell themselves: finding buyers is a skilled craft, and still does not form an important part of most foresters' training.

Does the commercial future for plantation forestry depend

more on the growing of special timbers? Even this is a gamble.
Hybrid poplars were planted in every valley bottom in eastern
England in the 1950s, in the expectation that they would be
valuable for making matches. Fast-growing though they were,
by the time they had grown the matchmakers no longer wanted
them. Ten years ago cherry was making big money as a furniture timber, and people were planting it for future reward.
Whether it will still be a fashionable furniture timber by the
time the trees have grown remains to be seen; anyone planting it
must look apprehensively at the Allegheny National Forest in
Pennsylvania, where there is far more cherry than in the whole
of England. Ultimately, the success of plantation forestry
depends less on growing timber than on marketing it.

Never again (at least among people with any trace of
common sense) will the growing of trees, or allowing trees to
grow, be presented in merely financial terms. This is not to say
that woodland economics are unimportant, but they have
become unpredictable, indeed unquantifiable. It is inherent in
the nature of plantation forestry that trees are grown to supply
the industries of forty years ago. It is seldom even possible to
know retrospectively whether a plantation did, in fact, make a
profit on the original investment: to preserve and interpret the
documents needed to find this out would require a rare
combination of archivist's and accountant's skills.

Many people plant trees in deliberate imitation of natural
woods. This is not a new development: it is a reversion to an
eighteenth-century practice, before conifers became fashionable. However, such new woods can, at best, be only a pastiche
of natural woodland. It is not too difficult to establish trees sim-
ulating the composition, if not the structure, of some sort of
ancient woodland. Whether the trees will copy the particular
ancient woodland that would have grown on the site had it remained woodland can never be known. It is more difficult to reproduce the herbaceous plants. Woods tend to be on poor soils
and are adapted to them: it is problematic to make a bluebell
wood out of farmland that has been drenched with fertiliser

over the last 150 years. New plantations should, therefore, be valued in their own right: they are not a replacement for the ancient woods destroyed, nor should they form a substitute for conserving the remaining ancient woods.

Ancient woods

Now that the big battalions of agriculture and modern forestry are in full retreat, the chief threat to ancient woodland is browsing animals, especially deer. Deer eat the tree foliage within reach and it does not grow again at a height at which they can get at it. They eat most woodland herbs except bracken and a few grasses. Deer are not dependent on the wood for their food: they eat everything available in the wood, and when that is gone they have plenty to eat in surrounding farmland. The result is a sadly truncated ecosystem. Instead of a wood with plenty of low cover, low-nesting birds, and a rich herbaceous flora, deer create a 'bottomless' wood, in which the trees' foliage stops abruptly at a browse-line limited by the height a deer can reach, no low cover, and ground vegetation limited to grasses and distasteful plants.

The problem is not the deer as such but *artificially large numbers* and *exotic species* of deer. In wildwood times there were two species of deer, red and roe. In the eleventh century fallow deer were introduced, ultimately from the Middle East. These three did not roam the country at large but were confined to parks and royal Forests, where special arrangements were made, such as pollarding, to combine trees and browsing animals. Ordinary woodland and woodmanship developed for a thousand years in the absence of deer, and their ecosystems do not withstand the reappearance of large herbivores.

In the twentieth century deer proliferated, and further oriental species escaped from parks. There are now more deer running around England than there have been for a thousand years, and more species than there have ever been: seven species within thirty miles of Cambridge, which is comparatively poor deer country. Chinese muntjac has become the most common deer, penetrating almost to the middle of Cambridge. The modern countryside is excellent deer habitat, with intermingled farmland and woods, little disturbance from people, few armed visitors, and plenty of autumn-sown cereals to keep deer alive over the winter. Deer are getting out of hand in many other countries too.

In mountain areas sheep have similar effects to deer, and fencing them out is the first step in woodland conservation. If sheep farming declines this will benefit woodland, at least until the sheep are replaced by more deer or feral goats.

What may happen and what can be done about it?

The cultural landscape, in particular trees and woods, involves living things. Trees are not 'Environment' – part of the scenery of the theatre of landscape; they are actors in the play, and each has its own separate part that needs to be understood.

Nearly everyone claims to be in favour of woodland, but in a vague and generic way. Even during the years of destruction myriads of trees were planted in the 'Plant a Tree in '73' campaign. People have put huge efforts into planting trees, but without asking whether previous phases of tree planting achieved their objectives. What happened to all those trees of 1973?

As far as one can predict for the next twenty years, woodland is likely to increase in quantity but deteriorate in quality. New woods will spring up naturally on abandoned land; people will make new plantations, and some of these plantations will probably turn, through neglect, into new natural woods. In general these will not be the same as existing woods. Their ground vegetation will not be bluebells and primroses, but the nettles and brambles of fertiliser-sodden soil. They will not, therefore, be a replacement for the ancient woods destroyed between 1950 and 1970.

The new woods of the future will be dominated by oak, ash, and birch – species that invade easily; they will not revert to the great limewoods of the prehistoric wildwood. Their development will be permanently altered by the introduction of the grey squirrel, which (among other effects) prevents hazel from recovering lost ground. Deer, although they often cannot prevent new woodland from arising, can prevent it from acquiring the structure or the full range of plants of natural woodland. By the mercy of God, none of the invasive plants is flammable: all over southern Europe one sees the tremendous consequences of allowing the countryside to be taken over by fire-promoting and fire-adapted species such as Aleppo pine and *Cistus* undershrubs.

The countryside will still have its remaining ancient woods, and will still be strewn with relics of former phases in the history of plantation forestry. The latter may gradually gain in individuality as time passes and they become more closely assimilated to natural woodland – to the extent that deer allow new trees and herbaceous plants to become established. Ancient woods will fare badly as deer continue to subtract from their flora and interfere with management, especially coppicing.

What can people do about this? There is plenty of advice available to landowners wanting to incorporate new values into land management. (Although not on every aspect: books on tree planting always contain lists of acceptable trees, but seldom advise the uninitiated reader how to choose which trees to plant out of a long list.) Many of the problems can be solved on a small scale. Phosphate can be removed from soil by growing a crop of nettles and cutting and removing them, and deer can be fenced out of areas of up to 50 acres (20 hectares) or so.

Should these solutions be extended to a landscape scale, and if so, how? It is remarkably difficult to get a group of neighbouring landowners to agree on anything more controversial than controlling rabbits: even managing a shared deer population is usually beyond them. One often hears the proposal 'Bring back the wolf': but there is little evidence that predators would really reduce the numbers of deer to reasonable levels, least of all in a countryside with a dense human population.

It should be obvious that governments are not competent bodies to handle rural and ecological matters. They are too short-lived to see through a policy until it takes effect, and they have a hard nose for bad, rather than good, scientific advice. The recent history of southern Europe is filled with comedies and tragedies resulting from inexperienced ministers applying pseudo-science to problems of agricultural retreat. British governments, after a long string of performances that rivalled those of their European neighbours, have won the World Cup with the combination of authoritarianism and ineffectiveness that declared foot-and-mouth disease to be an affair of state and then failed to control it.

The time has come to ask what are the cultural, biological, and maybe environmental values attached to trees and wood-land. Are these values already satisfied by the trees and woods that now exist, or is more woodland needed to fulfil them? Can such values be replicated by plantations? Or by woods that have arisen naturally on abandoned land? The answer may be No because they depend on passage of time – new woods, natural or artificial, cannot develop old gnarled coppice stools. Or because existing woods have been shaped by historic processes that cannot now be repeated. In the same way, a Victorian imitation of a medieval church, however thorough, never replicates the details of materials, patina, and weathering of the original.

Many of the values of trees are specific to old trees. Old trees of the future will not be created merely by planting new trees now: there are plenty of young trees, and a few more will make little difference. Instead old trees of the future will come from preserving the present middle-aged trees, and from starting new pollards among the present young trees. Pollarding enhances the character and habitats of old trees, and by making trees unsuit-able for timber may also deter future human generations from cutting them down.

Many features of the landscape, especially trees and wood-land, are the result of human neglect as well as human action.

Neglect is not a human failing to be lived down and forgotten in the hope that it will not recur. It is a recurrent aspect of human nature to be understood and provided for.

If my prediction of agricultural retreat is fulfilled, what will be the result? Parallels in other countries suggest a mosaic of natural recent woods and plantations, embedded in which will be the existing woods and trees. The woods will vary locally according to when and how each particular farm was abandoned, what the previous land-use was, what trees were already there, and so on.

Should bodies like the Countryside Agency, the Council for the Protection of Rural England, the Wildlife Trusts, and the National Trust try to influence this process? I hesitate to say they should: there is too little to go on. There has been no debate on what a more vegetated landscape should contain: on what would constitute 'enhancement' of the future countryside, compared with the results of letting events take their course. If I were Emperor of Northumbria I would have to command Northumbrians to do this, that, and the other; but I would have little confidence that posterity would uphold my decisions.

Part Three

National Attitudes

On the other side of sorrow: regenerating the Scottish Highlands and Islands

James Hunter

Many Highland glens, straths or valleys, together with many parts of the Hebrides, contain fewer people today than they did 5,000 years ago, in Neolithic times. Given the congested condition of much of the United Kingdom, this is an extraordinary state of affairs. It is all the more extraordinary in that many of the unpopulated localities now so common in the Highlands and Islands became unpopulated as a result of human action taken as recently as the nineteenth century. Prior to that point, to be sure, the localities in question were frequently subject to change, to crisis and, occasionally, to catastrophe. But although the human presence in them was one that altered both in size and composition over time, this presence was, in general terms, continuous. Picts may have given way to Gaels. Gaels may have given way, for a period at least, to Vikings. One language, one religion, one culture may have been displaced by another. Broadly speaking, however, the places I have in mind were spots where each of 200 or more human generations followed its predecessor until, around or just after 1800, succession of this sort was abruptly curtailed.

Hence the upset that can be caused to the people of the Highlands and Islands today by some of the language applied to their landscapes by environmentalists, conservationists and their allies. Especially in comparison with the densely populated regions a hundred or so miles to the south of it, an unpeopled Highland glen may look like wilderness or wild land. But insofar as this sort of vocabulary suggests or implies the presence of landscapes or ecosystems that have acquired their contemporary character by wholly natural means, it is misapplied in most Highlands and Islands contexts. These landscapes are stereotypically devoid of humanity because human communities of the sort associated with them for five millennia or more were removed from them – often in circumstances of great brutality – just two or three lifetimes ago.

If you have an eye for such things, it is possible to pick out, in any one of scores of nowadays-deserted Highland glens, what remains of the homes once occupied by that glen's former inhabitants. Since those homes were seldom very substantial to start with, and since their walls consisted mostly of locally gathered stone and turf, whole villages and townships have merged – in the space of the last two centuries – with their surroundings. Where once there were buildings, now there are only rectangular undulations in the grass, heather or bracken. Sometimes the names of emptied settlements have been lost. Often those names, where they survive, survive only in the heads of such elderly tradition-bearers as continue to take an interest in such matters – with even the most detailed of modern maps taking no cognisance of place names that were formerly familiar to very many people. But by walking the ground, and especially by keeping a sharp look-out for the telltale patches of the greener-than-average growth that denote abandoned cultivation, it is still possible to find – amid what appears to be the wildest of wild land – the traces of dozens, indeed hundreds, of abandoned habitations. To sit among any given set of such habitations, thinking of all the lives played out in and around them, is, if you have any sense of history, a deeply moving experience.

Repeopling the Highlands and Islands

This is not the place to explore the causes of the Highland Clearances – the name customarily given to the evictions and expropriations that culminated in numerous communities being, in effect, obliterated. But anyone seeking to understand the frequently conflicting emotions engendered by debate and discussion about landscape-related issues in the Highlands and Islands would do well to take heed of the fact that the Clearances, irrespective of what brought them about, still loom large in the collective psychology of those of us who live today in the northern half of Scotland. While few of us now contend that each and every emptied piece of territory ought eventually to be occupied again, we are the inheritors of a long-standing conviction that it is legitimate to aspire to, and plan for, a Highlands and Islands future involving the repeopling of at least some of those of our landscapes that are currently bereft of settlement.

The dream of a repopulated Highlands and Islands is one that has had plenty of adherents inside the Highlands and Islands themselves since the impact of eviction and ejection first began to be felt. Until recent times, however, this dream seemed – to most external observers at any rate – to have practically no chance of being realised. The Highland Clearances, in the sense of landlord-organised dispossessions of tenant after tenant, ceased during the 1880s, when popular protest led to the United Kingdom parliament granting security of tenure to such Highlands and Islands families as had somehow managed to hang on to a few acres of land. But despite legislative interventions of this kind, and despite a growing readiness on the part of successive United Kingdom governments to implement measures intended to stimulate and expand the region's economy, the area remained a deeply depressed one for much of the twentieth century. Prospects here were poor. Opportunity here was lacking. People – young people especially – reacted by leaving. Long after the Clearances ended, therefore, Highlands and

Islands depopulation remained endemic and, for long enough, things seemed set to continue indefinitely in this vein.

Today this has changed, and changed radically. In the course of the last 30 or so years – a period, incidentally, when the population of Scotland as a whole has been at best static – the population of the Highlands and Islands has risen by nearly 20 per cent. During those same 30 or so years, the number of people in employment in the region has increased by more than 40 per cent. While many of those gains have accrued to Inverness (one of the fastest-growing towns in the United Kingdom) and to urbanising localities in Inverness's immediate vicinity, parts of the Highlands and Islands periphery have also benefited. Take the case of the Isle of Skye, for example. At its height, in the 1840s, Skye's population stood at 24,000. Over the next twelve decades, the island's population fell year on year to just over 6,000 in the 1960s. Now it is in the neighbourhood of 10,000 once again, with the island's villages and townships filling, in consequence, with new faces, new families, new homes.

Admittedly, the Skye experience has been by no means universal. At the start of the twenty-first century there continue to be substantial segments of the Highlands and Islands – Kintyre, some of the Argyll islands, the Western Isles, parts of Caithness and Sutherland, the more outlying islands of both Orkney and Shetland – where economies are still shrinking and where populations are still falling. But the lack of progress so far in those places does not detract from the significance of what has been accomplished both in Skye and in several comparable districts. The fact that such districts have had their fortunes so spectacularly reversed is a remarkable achievement. It is an achievement, I am convinced, that can be replicated across the Highlands and Islands as a whole.

No single factor accounts for the undoubted advances made in the Highlands and Islands during recent years. Contributing hugely to those advances, however, is the fact that Highlands and Islands people are presently more positive about their background, their heritage and their surroundings than once they were. Where previously the Gaelic language was derided by an anglicising political establishment and rejected by those Highlands and Islands parents who thought it a barrier to their children's advancement, a new pride is now evident both in Gaelic itself and in the distinctive culture to which Gaelic gives access. Much the same is true of the musical traditions of the Highlands and Islands. And it is true, too, of landscape. While our mountains, our moorlands, our lochs, our coastlines and our woodlands were highly valued – as is demonstrated by many Gaelic songs and poems – in the relatively distant past, there were times in the twentieth century when spokespersons for the Highlands and Islands tended to be dismissive of claims that landscape deserved to be valued for its own sake. 'Scenery's fine, but you can't live on scenery,' people would repeat.

Such comments are heard less frequently today. Because of the growing importance attached to environmental issues all around the earth, it is beginning to be apparent to inhabitants of the Highlands and Islands that, in our landscape, we possess an asset with real scarcity value. This is a novel experience for us. For ages we tended to assume that the good things of life were, almost by definition, to be found in distant cities rather than in our own backyards. Now we realise that we possess something – this self-same scenery we were formerly in danger of devaluing and deriding – that others, city folk included, would dearly like to have. This has greatly helped to enhance our collective self-assurance.

Of equal significance in this regard has been the extent to which technological advances, particularly in the overlapping fields of telecommunications and computing, have made it possible to conduct today in a Highlands and Islands setting business activities of a type that would formerly have had to be carried out in faraway urban centres. One outcome has been the gradual emergence, not least in Skye, of rural economies that are much more diverse, and that possess far more potential for expansion, than the predominantly agricultural economies traditionally associated with the Highlands and Islands countryside.

Among the beneficiaries of this development are folk who, a generation or two back, would almost certainly have quit the Highlands and Islands on leaving school. Among those beneficiaries, too, are the many people who have recently been moving into the region, some from the south of Scotland, others from England. Because there are far more income-earning opportunities in the modern Highlands and Islands than there were 30 or more years ago, those people now have a reasonable chance of making a go of things in the localities in which they have settled. In numerous instances, however, their original decision to come north – and thus to contribute to the regeneration of the Highlands and Islands – can be traced to a desire to live among some of Europe's most outstanding landscapes.

In the Highlands and Islands, as in the rest of the United Kingdom, officialdom's instinctive response to rural repopulation has been to channel incoming households into suburbanised housing estates on the fringes of existing villages. Underlying such policies is an assumption that it ought to be a planning priority to protect and to safeguard our open and unbuilt-on land. But although such an assumption is perfectly understandable in relation to regions like the South East of England, where demand for housing seems little short of insatiable and where undeveloped countryside is in desperately short supply, it makes a lot less sense in the Highlands and Islands. Here, as I have emphasised, much of our countryside is less thickly populated than it was hundreds, even thousands, of years ago. And for better or worse, it is in the hope of setting up home in such countryside – with a view to having direct access, as it were, to our almost unrivalled natural heritage – that many of our in-migrants have elected to come to the Highlands and Islands in the first place.

The case for crofting

The prospect of enjoying some of Britain's most beautiful landscape explains the particular attraction to in-migrants of local-ities like the Isle of Skye. While Skye, in common with the rest of the Highlands and Islands, suffered from mass evictions during the nineteenth century, crofting (the peculiarly High-lands and Islands variant of smallholding) survived there into modern times. The typical croft is tiny in comparison with the average farm and so crofting districts contain many more houses than do districts given over to larger-scale agriculture. In prin-ciple, therefore, it is easier in a crofting area than in a farming area to obtain homes in the countryside. Moreover, homes in such areas tend to come with a croft, that is, with four, five or six acres (1.6-2.4 hectares) of land attached. This is something that appeals hugely to many of the folk who have been moving into the Highlands and Islands of late, not because those folk are intent on becoming subsistence farmers, but because they want, as country residents, to keep some animals, grow some vegeta-bles, plant some trees.

The crofters of this new breed are not departing in any fun-damental way from crofting as it was traditionally conceived. The minuscule acreages at their disposal made it impossible for crofters to be full-time agriculturalists, so they always combined the management of their crofts with other, non-agricultural, ac-tivities. In the past, a crofter might have been a fisherman, a tweed weaver, a tradesman of some sort. Today he or she may be a web-page designer, a teleworker, a graphic artist, a teacher, the operator of a tourism business. So heavily diversified is the total crofting economy in consequence, that most crofting house-holds in the Highlands and Islands depend on livestock rearing or crop growing for less than 10 per cent of their total income – something, of course, which makes crofting much less vulnera-ble than mainstream farming to crises of the kind that have re-cently been blighting the prospects of Britain's agricultural industry.

But if crofting, or something like it, offers a means of re-creating a vibrant countryside of the kind possessed by the Highlands and Islands prior to the Highland Clearances, its potential in this connection is limited by the fact that, as a result of the Clearances, the amount of land given over to crofting is much less than the amount devoted to those other land uses, such as hill sheep farming, that nineteenth-century landlords were seeking to promote when they emptied so many Highlands and Islands glens. Demand for crofts is well in excess of supply in those parts of the Highlands and Islands that have already experienced a degree of rural repopulation, such as Skye. Prices have duly been bid up to what are, by Highlands and Islands standards, extremely high levels. The ensuing exclusion of local people – the young in particular – from the market for crofts and, for that matter, from the housing market as a whole, is becoming a potent source of social discord. This is one reason why it is to be hoped that the land reform agenda being developed by the Scottish Executive, Scotland's devolved government, will permit and encourage the expansion of crofting.

The case for such a development is made all the stronger by the troubles afflicting hill farms of the type constituted in the course of the Highland Clearances. Those hill farms began by being extremely profitable and as such made perfect business sense from the point of view of the landlords responsible for their emergence. However, in the Highlands and Islands at any rate, hill farming has long since ceased to be viable in any free-market sense of the term, having been propped up, for the last half-century or more, by subsidies of one kind or another. Such subsidies are almost certain to be scaled back over the next ten or so years, as more and more countries in Central and Eastern Europe join the EU and as the EU's Common Agricultural Policy becomes (because of the pressures EU expansion is bound to place upon it) less generous. When that happens, the underlying unsustainability of hill farming of the current Highlands and Islands type will be starkly exposed. This is not to say that agriculture will vanish entirely from these hills. But it is to insist that hill farming is destined to contract, perhaps significantly.

Arguably, this contraction is already occurring, as can be seen from the virtual cessation of father-to-son succession. Hill farmers themselves, particularly tenant hill farmers who do not have the option of experimenting with alternative land uses, can see the writing all too visibly on the wall.

In other upland areas of the United Kingdom – in Wales and the Lake District, for instance – the growing, and potentially terminal, difficulties confronting hill farmers have led to calls for their role to be reassessed. Perhaps, it is said, hill farmers should be seen more as guardians of the landscape than as sheep producers. After all, it is added, if hill farmers and their flocks were to disappear from the Lake District then that locality's characteristic landscape – a landscape shaped largely by an agricultural system rooted in medieval times – would be subject to all sorts of alterations. Many people could well regard these alterations, which would inevitably include a steady expansion of the area under trees, as undesirable. It therefore follows that ways would have to be found of keeping Lake District hill farming in existence for reasons rooted in landscape conservation rather than in livestock rearing.

Similar arguments have been voiced in a Highlands and Islands context. But it would be a pity, I believe, if those arguments were to prevail at the expense of more imaginative, more innovative, possibilities.

Indubitably, Highlands and Islands hill farming has been in decline, continues to be in decline and seems set to become an even more problematic land use in the not-too-distant future. At the same time, however, more and more people want to, and can, create new livelihoods for themselves in the Highlands and Islands countryside, livelihoods that those same people frequently wish to pursue in conjunction with the occupation and management of a few acres of land. In those circumstances, or so it seems to me, it makes excellent sense to take some of the land presently in hill farming and turn it into crofts.

While this would be to reverse the Highland Clearances to a small extent at least, that should not in itself prevent us from embarking on such an exercise. The desire to right past wrongs is certainly one element in the thinking that underpins the politics of land reform in present-day Scotland, yet the cause of land reform would have no great momentum, and would deserve none, were it conceived solely as a means of inflicting some sort of posthumous revenge on the perpetrators of nineteenth-century crimes, however harsh and heartless those crimes may have been. Much more to the point is the contribution land reform might make, and is already making, to providing today's Highlands and Islands communities with powers and opportunities of the sort formerly monopolised, because of the exceptionally concentrated nature of land ownership in the Highlands and Islands, by a tiny number of landed proprietors.

Enabling rural communities to exercise more control over a resource as basic as land allows the communities in question to do things at their own hand, and for their own benefit, which would otherwise be impossible to accomplish. Developmentally, this is crucial. Communities condemned to perpetual dependence on frequently absentee landlords – no matter how benevolent or well intentioned – will forever lack confidence in their own capacities. Lacking such confidence, they will also lack hope, vision and entrepreneurialism, which are all qualities essential to their long-term progress.

Further initiatives

The desire to build self-confident communities underlies much of the thinking of the Highlands and Islands Enterprise (HIE), the regional development agency whose board I chair. HIE has actively assisted, for some years now, the spread of community ownership of land in the Highlands and Islands. As a result, islands like Eigg and Gigha, together with several substantial estates in Assynt, in Knoydart and on the Isle of Lewis, have been acquired by variously constituted local groupings – as have many other, usually smaller, pieces of territory. The land reform legislation being developed by the Scottish Executive is meant to encourage other communities to follow this lead. Meanwhile, a Land Fund, with £10 million to spend over its first two years, has been set up with a view to helping such communities find the necessary finance.

Experimentation of the type now occurring in relation to Highlands and Islands land ownership is never straightforward. This is most apparent in the problems that have beset HIE's own attempt to give practical expression to the notion that a part of the huge area of land presently in hill farming might be utilised to provide both homes and croft-type smallholdings. Central to our efforts in this regard is the Orbost estate on the Isle of Skye. This estate, previously a single hill farm, was bought by HIE in 1998 with a view to fostering the emergence of a new community. At Orbost we wanted to give people the chance to combine the occupancy of a piece of land with the tenancy of a home made available at an affordable rent. This is critically important if we are to make it possible for the majority of young people to live, work and develop businesses in the Highlands and Islands countryside, where wages tend to be low and housing costs high.

The Orbost initiative is beginning to yield some positive results, not least by providing a number of Skye folk with the sort of opportunities they otherwise would never have had. Along the way, however, local feeling has occasionally run very high, with sharp divisions emerging as to whether or not it makes sense to repopulate (at no small cost to the public purse) a locality where, from the time that most of the Orbost estate was cleared of its original inhabitants in the early nineteenth century, human beings have been thin on the ground.

Controversy of this sort notwithstanding, such repopulation schemes will continue. Already, for example, two further sets of newly created smallholdings are taking shape at Balmacara and Fernaig on the Highland mainland adjacent to Skye. The high level of demand for these smallholdings shows that others are required. Suddenly, therefore, the reoccupation of places that have been either uninhabited or very scantily peopled since the era of the Clearances has ceased to be a long-run aspiration and has become, instead, something that is actually happening.

It is worth emphasising in this context just how thinly populated the Highlands and Islands have become. Even if the region's present population were to be doubled, something that certainly will not happen in the foreseeable future, the Highlands and Islands would still be the least densely populated part of the EU outside arctic and subarctic Scandinavia. Without in any way jeopardising those qualities that make the area so attractive to so many of its visitors, therefore, we can readily have a number of new rural settlements – particularly in localities such as Skye, Lochalsh and Lochaber, where demand for housing is high and where economies are growing fast.

All such new settlements will need both careful thought and careful handling. It would be good, for example, if we could ensure that new housing in the Highlands and Islands countryside makes as much use as possible of locally quarried stone, locally obtained timber and the like. It would be good, too, if new homes were architecturally in keeping with the building traditions of our area – just as it would clearly be desirable, by minimising new road building and by utilising woodland cover, to make new settlements every bit as visually unobtrusive as their eighteenth-century predecessors once were.

When we look at how the repopulation of some parts at least of the Highlands and Islands would proceed in practice, the mass of legislation concerning crofting and related matters might seem an obvious starting point. That legislation, however, was not designed to deal with current circumstances. The complex and somewhat cumbersome administrative structures surrounding crofting are predicated on the notion that crofters need defending from a whole series of threats to their existence. While this approach was necessary when crofting was indubitably at risk, whether from evicting landlords in the nineteenth century or deprivation-driven depopulation in the twentieth, it makes less sense now that crofting, or some variant of it, is beginning to have the potential to expand beyond its present boundaries.

The rules, regulations and bureaucracy that loom so large in relation to crofting have certainly had the effect of ensuring its survival into the twenty-first century, but they have also tended to make the crofting population unhealthily reliant on external agencies of one kind or another. This is at odds with what today's land reform is aiming to achieve. Hence the importance of ensuring that any new communities resulting from such reform are encouraged to be much more self-governing than crofting communities have traditionally been.

But if there are aspects of the crofting inheritance that deserved to be questioned, even rejected, there are others we would do well to nurture, especially if land reform and rural repopulation are to proceed in ways that safeguard, indeed enhance, the natural heritage of the Highlands and Islands.

Timeless connections

Celebration of our national heritage is central to much Gaelic literature. Also at the heart of a great deal of this literature — some of it very ancient — is the nature of the link between human communities and their surroundings. Historically in the Highlands and Islands, the connection between people and place was of immense significance. It surfaces over and over again in songs, poems, tales and stories. It underpinned, in addition, the concept of *duthchas*, a Gaelic term embodying the age-old conviction that simply to occupy a given piece of territory for several generations was, by virtue of that fact alone, to have established a right to inhabit the same piece of territory for all time coming. This sort of thinking, needless to say, was totally disregarded by the landlords who organised the Highland Clearances. But so strong were the bonds between particular communities and the localities in which they were situated that the breaking of those bonds in the course of the Clearances was, from a Highlands and Islands perspective, one of the more brutal features of a very brutal period.

This is why places once peopled but which are peopled no longer seem, in a Highlands and Islands context, incomplete. It also explains the tendency in a great deal of the modern literature of the region to look forward to a time when our deserted landscapes are made whole again by having human beings restored to them. This tendency is apparent in the output of several of the leading Highlands and Islands writers of the twentieth century, not least in the work of the late Sorley MacLean, who was, by common consent, easily the most eminent Gaelic poet of the modern era. MacLean, in the words of a fellow poet, Iain Crichton Smith, 'felt and sensed quite clearly the desolation, the sadness, the terrible emptiness of the Highlands, its ghosts and presences, in an absolute intuitional music'. But MacLean, for all that, never ceased to believe in the possibility of Highlands and Islands betterment. To him, the region remained capable of regeneration, rebirth, repopulation, despite its past sufferings. This is made explicit in the closing lines of his poem, *An Cuilithionn* ('The Cuillin'):

Thar bochdainn, caithimh, fiabhrais, àmhghair,
thar anacothrom, eucoir, ainneart, ànraidh,
thar truaighe, eu-dòchas, gamhlas, cuilbheart,
thar ciont is truaillidheachd; gu furachair,
gu treunmhor chithear an Cuilithionn
's e 'g éirigh air taobh eile duilghe.

Beyond poverty, consumption, fever, agony,
beyond hardship, wrong, tyranny, distress,
beyond misery, despair, hatred, treachery,
beyond guilt and defilement; watchful,
heroic, the Cuillin is seen
rising on the other side of sorrow.

Here the Cuillin, the sharply pinnacled mountains that tower so impressively over the Isle of Skye, become an emblem of salvation. They beckon us towards a future in which people become once more an integral component of Highlands and Islands landscapes. Getting that future right will not be easy. But from a Highlands and Islands standpoint it is infinitely more heartening to be contemplating how to cope with repopulation than with its opposite.

Heritage landscapes in Wales: cultural and economic regeneration

Richard Keen

Wales is a land of geological complexity with formations dating from the Pre-Cambrian period about 570 million years ago to the Jurassic about 190–150 million years ago. It has provided a virtual textbook for generations of scientists. Glacial action helped to shape the land, forming the classic steep-sided valleys of Snowdonia and the smaller but equally steep escarpments of Old Red Sandstone that give such distinction to the Brecon Beacons in the south. Rocks and minerals provided the basis for the industrialisation of Wales from prehistoric times, with Carboniferous Limestone and Coal Measures forming the foundation for economic exploitation from the eighteenth century onwards.

Rainfall is the most dominant feature of the climate, in the form of either short but intensive showers or prolonged but lighter precipitation. Cloud formations above the mountainous areas can be a visual spectacle and their shapes and sizes have inspired generations of writers and artists as well as influencing the local idiom. An understanding of the effect of landscape and weather is encapsulated in the local description 'esgobion Bangor yn eu gwenwisg' – 'the bishops of Bangor in their surplices', as the great white clouds gather from the west over the mountains.

More than half of Wales is above the 152-metre contour and

much of the rest is described as upland because of its climate and soil. Although parts of the country are perceived as industrial, it is agriculture and forestry that dominate the landscape. Even in 'industrial' southeast Wales the area actually covered by building and industry is only about 30 per cent of the land mass, and that is confined mainly to the coastal strip and floors and lower slopes of the valleys. Over 90 per cent of the land in Wales is designated as rural and out of that 82 per cent is used for agriculture, with forestry accounting for about 12 per cent.

It is a country of breathtaking visual and sensory contrast where the landscape and weather can change rapidly and over short distances. In the space of a few hours' travel the stark grandeur of Cwm Idwal in Snowdonia can be contrasted with the rounded hills around Aberystwyth, the sweep of St Brides Bay in Pembrokeshire, that 'little England beyond Wales',[1] or the 'ugly, lovely town'[2] of Swansea now undergoing a mini-renaissance.

It is a land of economic contrasts as well. Along parts of the M4 and A55 corridors income levels can, in pockets, equal in real terms those of south-east England. However, within relatively short distances of those main roads the picture can change rapidly, taking you through places such as Blaenau Gwent and Blaenau Ffestiniog, which are among the most economically deprived areas in Britain. In general economic terms Wales lags behind England.

The landscape has influenced and has been influenced by human activity to a greater or lesser extent since the retreat of the last glaciers some 10,000 years ago, although the earliest traces of human life have been dated to around 225000 BC. The landscape is rich with the remains of history, from prehistoric hill forts, Roman roads and settlements, the towns, castles and abbeys of the Middle Ages, great houses and estates to the very visible impact of industrialisation.

Tree cover up to a level of approximately 610 metres was reduced from the Neolithic period onwards as felling, cultivation and grazing changed the nature of the vegetation. Subsequent agricultural practices have continued this trend and to a large extent grassland has replaced woodland. Soils in the uplands are generally thin and the low provision rate of organic material has not led to much improvement. Seasonal stock movement was, and still is, necessary to exploit the limited sources of good-quality grazing.

Seasonal movement of livestock has therefore been central to the successful operation of an agricultural economy. In 1786, Thomas Pennant described how this system worked in Snowdonia, before the wider spread of enclosures:

> Its produce is cattle and sheep which, during summer, keep very high in the mountains followed by the owners with their families, who reside in that season in their hafodtai or summer dairy-houses . . . Towards winter, they descend to their hen dref, or old dwelling . . . [3]

Similar conditions prevail today, with the need for overwintering pastures being an important consideration for the upland farmer. It is often an economic imperative that land ownership and tenancy include areas of both upland and lowland.

Wales has been a land of tenanted farms, where the administration of the large estates, often owned by absentee landlords, was left in the control of locally based agents. With the decline in the larger estates, the number of owner-occupied farms increased, but this was a hard-won right both in economic and legal terms. It is therefore not surprising that a sense of ownership can be a powerful influence in strengthening the relationship of families with their particular farm.

The spread of the railways in the second half of the nineteenth century saw the demise of the drover – the economic lifeline for many agricultural communities from the eighteenth century onwards – and was part of the important system of communication that enabled large-scale exploitation of natural resources to take place. Railways provided a network of communications across the country, which existed until the widespread rail closures of the 1960s.

Metals have been mined in Wales since prehistoric times, as excavation at the copper mines at the Great Orme near Llandudno, at Parys Mountain on Anglesey and at Copa Hill near Aberystwyth has proved. No part of the country has been ignored in the search for metals throughout history.

The landscape of northwest Wales is, in parts, dominated by traces of the slate industry, ranging from small, scattered quarries to the massive undertakings of the Dinorwig and Penrhyn quarries. Slate quarrying is a good example of an industry set in a rural landscape. The working communities of many of the quarries operated within the dual economies of industry and agriculture. Blaenau Ffestiniog, along with Llanberis and Bethesda, are larger-scale industrial settlements that grew in parallel with the expansion of the industry during the nineteenth century. Lack of development during the second half of the twentieth century has meant that their character has not changed much. The slate industry has been described as the most Welsh of Welsh industries and the quarrying communities were radical, literate and cultured, with an intense sense of place and pride in their skills, language and traditions. The quarries, their tips and the network of narrow-gauge railways that linked the remote places to the small ports are a remarkable testament to their endeavour.

The valleys of south Wales and the area around Wrexham in the north-east were the most heavily industrialised regions. Iron was manufactured first in small charcoal-fuelled furnaces and later in much larger coke-fuelled furnaces. The iron industry formed a solid foundation for the development of steel and tin-plate manufacture. Coal mining, which initially served the iron industry, grew to become the principal industry of Wales. Industrialisation had a profound and lasting effect on the culture of the country as the movement of farming people from Wales and beyond into industrial settlements began to take place on a massive scale. After 1875[4] the mining communities expanded at a hitherto unparalleled rate, reaching their peak immediately before the First World War, after which they began their slow and painful decline, culminating in their near-obliteration in the 1980s.

Cardiff, Swansea and their neighbouring ports served the industrial hinterlands. Both cities are currently undergoing a substantial revival in their fortunes. Dockland development in Cardiff since the 1980s has transformed both the appearance and culture of the formerly depressed area and the creation of a maritime quarter in Swansea heralded a number of similar enterprises elsewhere, while the development of the city of Swansea itself continues.

Previous reliance on monolithic heavy industry, accompanied by a lack of investment and diversification, resulted in the economic and social trauma of the last quarter of the twentieth century, as many of the industrial communities came to terms with large-scale deprivation and redundancy. Closure of a pit or steelworks has more than an economic effect on a community – it takes away the very focus of life and often results in a lowering of cultural self-esteem. The inward investment that was so vital to help compensate for the losses incurred by the reduction in heavy industries is now under pressure. The vital requirement is the generation of high-value jobs and a broadening of the entrepreneurial culture.

A cultural understanding of landscape

The relationship between landscape and employment is becoming increasingly relevant, as quality of life is an important consideration when companies are looking at the possibilities of relocation. All too often, however, development in the countryside results in the transference of urban design into a rural environment with little regard for local building materials, let alone the local vernacular. Wales does not have a good record in this regard and there have been visual travesties that would never have been allowed, for instance, in the Cotswolds of England. The 'imperative of the aesthetic'[5] must be allowed to influence decision-making in Wales as well. It is also important to ensure that new landscapes – those that form part of a development – enrich biodiversity, help conserve cultural identity and have regard to local variations in the landscape.

LANDMAP (Landscape Assessment and Decision Making Process), introduced and operated by the Countryside Council for Wales in conjunction with local authorities and National Park authorities, is providing a wider view of the landscape. The draft handbook[6] defines landscape as 'the environment perceived, predominantly visually but also with all the other senses. Sight, smell, feel and sound all contribute to landscape appreciation. Our experience of landscape is also affected by cultural background and personal and professional interests.'

Perceptions of landscape depend upon time and place and upon the person making them. Judgements on what should or should not be retained in a landscape are part of a long-standing conditioning that can be traced back to the eighteenth and nineteenth centuries. Dr Johnson thought that Wales 'held nothing for the speculation of the traveller' and a certain E.B. described Snowdon as 'the fag end of creation, the very rubbish of Noah's Flood'.[7]

Such views are in sharp contrast to today's appreciation of Snowdon as the most significant and well-known area of landscape in Wales. It occupies an important place in the hearts and minds of many and is the most visited mountain in Britain, with some 400,000 visitors annually. Culturally it has played a fundamental role in helping to formulate our present understanding of beauty and conservation.

Snowdon is perhaps a classic place to ask the question 'Whose landscape is it anyway?' Does it 'belong' to the city dweller in need of recreation or does it 'belong' to the farmer or, more particularly, the farm tenant? It is a mountain that is seen by some in the area as being the 'English' mountain. Indeed, the concept of the National Park rooted in the Picturesque is viewed in the same light. This dichotomy between local and external perceptions of ownership can be extended across the whole of Britain.

There is now substantial economic dependence upon a cultural understanding of landscape that can place significance on one particular location, yet allow another area to suffer extremes of exploitation. Tintern Abbey and the Wye Valley are examples of the persuasive powers of antiquarianism and the Picturesque. Such has been the impact of the 'Wye Tour' on the collective consciousness that it has contributed a great deal to the present understanding of landscape. The Wye Valley, and Tintern in particular, is one of the most popular visitor attractions in south Wales.

Yet there are other places that represent a different kind of cultural understanding. They may, at times, be equally busy but they are not viewed today with the same regard. The serried ranks of caravans on the outskirts of Porthcawl at Trecco Bay were, and perhaps still are, part of the biggest caravan park in Europe, complete with its own entertainment facilities and shopping complex. Trecco Bay may be seen as something of an anomaly in these days of cheap air flights to guaranteed sunshine but it served a very important role in the years immediately after the Second World War. The caravans were mainly occupied by coalfield workers and their families and by the 1960s were providing accommodation for over 16,000 people in nearly 3,500 caravans. The similarity of rows of caravans to the terraced rows of miners' cottages in the valleys of south Wales is all too apparent. Perhaps Trecco Bay was a duplication of the close-knit community and culture of a Rhondda or an Ebbw Vale. The caravan parks along the north Wales coast served a similar purpose for the working communities of Liverpool and Manchester.

A hierarchy of landscape

Determining what is significant is therefore crucial. The ranking of landscape is important and has become increasingly so over the past fifty years, as the number of designations has expanded. Assessments are already in place via Sites of Special Scientific Interest, Areas of Outstanding Natural Beauty and so on. The designation of a National Park is part of this assessment and when the National Trust preserves an area of landscape that too represents an assessment of significance. Added to these are increasing numbers of European designations under the EU Habitats Directive that are continuing the trend towards centralised designation.

About 25 per cent of the land of Wales is designated as National Park or as an Area of Outstanding Natural Beauty, and approximately one-third of the coastline as Heritage Coast. There are 1,005 Sites of Special Scientific Interest, with proposals to add a further 400. There has been an increase in the number of Conservation Areas and Cadw: Welsh Historic Monuments has listed some 25,000 buildings as being of architectural or historic importance.

Our statutory landscapes affect perceptions as to which area or place is more important than another. The boundary of a National Park can confer significance on one side of a mountain or stretch of coastline but not on the other. Yet it is the same mountain or the same cliff. Perhaps the best-known example is the exclusion of Blaenau Ffestiniog from the Snowdonia National Park in 1951. The decision to do so may have been an expression of the way industry and industrial settlements were perceived, yet in landscape terms it can be argued that the towering, but carefully constructed, tips of slate waste are heroic in their scale and have a significance that can be compared to the scattered ruins of classical antiquity.

It may be that this difference in understanding exists in the heart as well as the mind, influenced by formal education and received values. Landscape can be a construction of inherited values and traditions and reflect the local geology and geomorphology where, largely because of technological limitations, human intervention has been in harmony with environment. It can be a place – the *bro*, the *cwm*, the river, the woodland – holding great significance for a community that cherishes the familiar but which is ignored or considered commonplace by the external specialist.

The late nineteenth- and early twentieth-century historian, Sir O. M. Edwards, described landscape from a Welsh perspective: 'Our land is a living thing, not a grave of forgetfulness under our feet. Every hill has its history, every locality its romance, every part of the landscape wears its particular glory.'

This is evident in the place names of Wales, rich in landscape imagery and history, such as Esgair Cywion – the ridge of the young horses, Rhosllannerchrugog – moorland of the heather glade, and Twyn-y-gwynt – hillock of the wind. The cultural enrichment provided by two languages contributes much to the understanding of landscape. In fact the language has a resonance that transcends place names. It is a fundamental part of the character of Wales and although spoken by a minority has powerful relevance across the land. There has been a small but significant increase in the number of people speaking Welsh, arising from the influence of the Welsh schools, where the curriculum is taught through the medium of Welsh. This has happened, surprisingly, mainly in the anglicised urban areas of the south-east, rather than in the traditional Welsh-speaking heartland of north and west Wales, where use of the language continues to decline.

Community involvement in history and preservation is vital. An example of its importance is to be found in the area bordering Rhondda, within the parish of Llanwonno. The landscape has strong associations with the astonishing athletic feats of Griffith Morgan, or Guto Nyth Bran (1700–37), who, among other deeds, was reputed to have raced and caught a hare. There is a strong local appreciation of the man, his deeds – be they fact or fiction – and his landscape, to such an extent that there is now a growing identification of an area of countryside as Guto Nyth Bran Land.

And why not? Wales already has Kilvert's Country, Cordell Country and the Dylan Thomas Trail. It is only a matter of time before R. S. Thomas Land appears on the signposts. It is certainly an irony that visitors to Pontypridd seek out the birthplace of

Tom Jones without giving a nod of recognition to the plaque commemorating the place where the national anthem of Wales was written.

Despite the commercialisation of such associations, landscape is increasingly being perceived as having great historic significance. The publication of registers of landscapes of historic interest demonstrates that this understanding is not confined to a designated site or particular location but can reach across large areas. Although the registers[8] are advisory, non-statutory instruments, they have already achieved much in raising awareness and providing information about parts of the country that have not previously been accorded heritage importance. An example is the entry for the Nantlle Valley, which has been included for its importance as a slate quarrying area, for its connections with the medieval Welsh tales, the *Mabinogi*, and for its significance in the nineteenth-century dispute 'between the "Diluvialists", who believed in the Biblical flood, and the "Glacialists", who supported the Glacial Theory'.[9]

The layered hedge of the Vale of Glamorgan, the slate pillar fence of Gwynedd, the grouted roof of a Pembrokeshire cottage and the timber-framed building of Radnorshire are visible expressions of a close relationship between the people and their particular places. The challenge is to retain these important landscape characteristics.

Reform and reclamation

Wales has undergone the same developments as other parts of Britain. Everywhere can be found the same garage forecourt, the same fast-food outlet sign, the same plastic windows and the spread of the cul-de-sac culture. There are the pedestrian schemes where the inevitable brick paving and cast-iron bollards appear as instant 'heritage' ordered from a catalogue. In the countryside every new scheme seems to be accompanied by standardised road splays and lighting.

There is, however, evidence of a deepening understanding of local character in the attempts to introduce new industrial and business buildings that either are architectural statements in their own right or perhaps make some attempt to match the local character. All too often, however, economics takes precedence and anonymous 'crinkly sheds' appear. When trading estates were developed immediately after the Second World War, to offset the desperate unemployment in the valleys of south Wales, attempts were made to introduce good standards of design. There is no reason why places of employment should not be places of visual enjoyment as well. The estates also had a profound social effect in that they provided employment for female labour, a trend that continues to this day.

Many newer industrial estates and business parks have appeared on the sites of former industries. The reclaiming of these sites has been part of one of the largest programmes of landscape re-forming in Europe. The reviled colliery waste tips, once treated with ambivalence, became symbols of exploitation and horror after the disaster in Aberfan on 21 October 1966. The few now remaining are hidden away on mountain tops but could, subject to stringent safety checks, be left untouched as a reminder of the mining industry.

Since reclamation work began in the 1970s thousands of acres of derelict land have been reclaimed and perhaps it is not without significance that in south Wales this was called 'Greening the Valleys'.[10] Many earlier reclamation schemes gave little or no thought to the visual quality of the new landscapes being created. The green-grassed, graded slopes punctuated by concrete spillways bore little relationship to their immediate natural environment.

The same period has seen the spread of country parks, where many areas of derelict land have been converted into accessible safe areas. This has resulted in an increased awareness of environmental issues and a sense of pride in the landscape. The Dare Valley Park near Aberdare, opened in 1974, was the first of its kind in Britain.

Another aspect of reclamation has been the loss of the

industrial heritage of Wales, so often portrayed as a reminder of a past best forgotten, a situation somewhat offset by the designation in 2000 of Blaenafon and its surrounding landscape as an area with World Heritage Status. This has marked a new awareness of the importance of industrial landscapes. The loss of some of the seminal buildings in the 1970s and 1980s – the Gilchrist Thomas laboratory in Blaenafon and the William Siemens open hearth furnaces in Swansea – is regrettable since such remains represented so many aspects of industrialisation, both technological and cultural. The challenge now lies in finding new uses for old buildings when they may not be readily adaptable. The demolition of the twentieth-century Dunlop Semtex factory in Brynmawr in 2001 was particularly unfortunate in this respect as it could easily have been used for new purposes.

There has been a distinct change in the appreciation of what constitutes the heritage of Wales, as more aspects are being included in the preservation 'portfolio'. There are still places, however, that are not being catered for, since they do not fall into the well-established categories of archaeological and architectural interest. A strategy is needed that looks at the future of the past and its landscapes, and that takes account of what is valued by local communities. Such a survey would reveal the gaps and suggest how they might be filled.

Farming and landscape

It is within the agricultural landscapes of Wales that the greatest pressures are now being felt. Even before the devastating effects of foot-and-mouth disease, the agricultural industry in Wales had been under stress for over ten years, as prices fell, costs increased and subsidies were reduced. Net incomes for all dairy and livestock farms in Wales are forecast to increase to an average of £5,300 in 2001/02.[11] In the year 2000, farm incomes in some places were less than £5,000 per annum, a fall of about 40 per cent since 1996.

Farming used to amount to much more than food production; it supported a complex and interlocking culture where cooperation and communal activity were paramount. Much of the social interaction was directly related to the demands of the business. Modern agriculture, in contrast, is a solitary operation where the farm operates as an independent cost centre reliant more on machinery and less on manpower, and where many of the activities in the farming calendar are now performed by outside contractors. There is also a steady decline in family-run farms, with more farmers becoming part-timers who have salaried jobs elsewhere. Amalgamation of farms into larger agri-businesses is another continuing trend.

Many aspects of rural life are at risk, as the numbers of farms and those engaged in farming decline. In the European Task Force's *National Economic Development Strategy* the situation was described thus:

> The viability of rural communities, the traditional
> heartland of Welsh culture and language, is under threat
> from the depression in the agriculture industry – the
> economic core of the rural economy – from the demise
> of local goods and services such as the village shop and
> post office as well as public transport, health-care
> provision and recreational amenities.[12]

Even the natural meeting point of the local cattle market may disappear as buying and selling by electronic media develops. The abrupt cultural loss that was experienced by industrial workers when their colliery or steelworks closed is now being felt by the farming community, albeit in a slower and possibly more insidious manner. The suicide rate in farming families, particularly among men, is worrying and perhaps indicative of the pressure they are experiencing. Farming is a lonely occupation and for those who are part of this long-standing tradition it can be difficult to make a break from the land and find new occupations.

The introduction of subsidies under the Hill Farming Act of 1946, and the emphasis it placed on food production rather than economic sustainability, have been blamed for the changes that have been wrought on the land. The impact of the Act on wildlife has been devastating and has, according to the authors of *Silent Fields*,[13] resulted in the loss of 98 per cent of flower-rich meadows, a dramatic decline in bird, butterfly and moth species and the acidification of over 1,200 kilometres of upland streams. Alongside this has been the increase in sheep numbers from approximately 6 million in the early 1970s to over 11 million by the early 1990s, with a concomitant increase in overgrazing, ecological damage through drainage and the heavy application of fertiliser.

Welsh Office figures show that 53,000 people were engaged in the agricultural industry in 1996, about 10 per cent of the rural workforce. In 2001 that figure was estimated at just over 42,000. Agricultural output was put at £1,086 million in 1994. The same report[14] shows that in 1991 rural tourism employed a workforce of over 27,000 and generated £700 million per annum. In 1999 tourism spending associated with 'environment-motivated' trips amounted to £821 million. An analysis of the economic benefits that can be ascribed to the environment has estimated that £6 billion of GDP is directly dependent on the environment. This represents one in six jobs in Wales and a wage bill of about £1.8 billion.[15]

Stewardship of the countryside, once part of farming life, now demands specific plans. For this reason the Countryside Council for Wales has developed and run schemes directed at protecting local character and caring for essential habitats. The most important of these, Tir Cymen, was launched in 1992 in three pilot areas. Under the scheme, farmers are paid a premium to resume their traditional roles as custodians of the countryside. As such, Tir Cymen represents a shift away from production subsidies or 'headage payments' towards payment for 'environmental goods'. These environmental payments cover the conservation of wildlife habitats, the retention of important

landscape features including traditional buildings and boundaries, archaeological and historic features, and the provision of
greater public access. Tir Gofal, part of the Tir Cymen project,
extends the system to the rest of Wales, as part of an integrated
agri-environment scheme.

The long-term success of the schemes remains to be seen.
They run for ten years, are voluntary and have little or nothing
to offer the larger farmers and those who are well advanced
with modern farming methods. To qualify for Tir Gofal requires
existing biodiversity and heritage capital. Where, for instance,
field boundaries have been removed and land drained, the level
of grant available is simply not sufficient to make a return to Tir
Gofal requirements economically viable. There is also the
counterargument that these schemes and other farming support
mechanisms are symptoms of a deep-rooted sentimentality
about the countryside and that farming, as has already happened
with industry, should be exposed to the forces of the open
market and, if necessary, the countryside left to change naturally
without intervention.

Tir Cymen and other similar activities have begun to tease
out a way forward whereby the countryside – scenery might be
a better description – can be cared for in a way that is compati-
ble with the demands of urban and rural dwellers alike. Who
better than farmers and landowners to undertake the work, and
what better than the public purse to pay for it? This is subsidy by
another name but one that has greater benefit for the wider
community. If it can be linked to a farming economy that produces more food for a local market and responds to the values of
the consumer, then there may be a future for the upland farmer.
The rise in consumption of organic produce is an encouraging
indicator, although more needs to be produced in Britain. Is it
too much to foresee a return to locally produced goods sold in
local shops and supermarkets? Could we even hope that some of
the larger supermarket chains might use local shops as outlets
for bulk-purchased goods?

Diversification is not new; the economic realities of life have already forced farmers into other activities and many farming families can exist only with secondary incomes. The success of any such objective cannot be achieved from an aesthetic or environmental perspective alone. It must have the support of the general public and be shown to have positive economic and social outcomes.

Prospects

Over the past thirty years there have been noticeable changes in the number and quality of heritage projects in Wales. The general trend is for this number to rise, though the shortage of funding makes long-term sustainability questionable. Larger 'flagship' projects can easily drain the coffers of the funding sources and those organisations sustained by government funding can make it difficult for smaller private schemes to compete in the visitor market. There are relatively few sources of finance available – the greater proportion of human and natural heritage funding comes from government via the local authorities and the state organisations, with the relative newcomer, the Heritage Lottery Fund, making a positive and valuable contribution. European funding is also important but the time constraints to which particular European programmes are subject tend to focus attention on capital projects. After the initial hype and publicity of capital-intensive projects there may follow, perhaps five or ten years later, when visitor figures have levelled out, difficulty in sustaining growth and paying salaries. There are examples of past schemes that are now in need of maintenance and are the responsibility of the already over-stretched local authorities. An assessment of long-term sustainability at the outset is crucial.

Partnerships and a greater cohesion between public and private activity are vital if we are to avoid such mistakes in the future. Achieving this should be a relatively straightforward process in Wales, where Welsh Assembly government now plays such a crucial role. In a country that has often been described as a 'large village' there are signs that such cohesion is already occurring. The remarkable achievement that resulted in Blaenafon being given World Heritage Status was the result of a successful partnership of a number of bodies working closely together for a common aim. Underlying everything at Blaenafon was the need for economic regeneration. This, together with the need for greater local representation and benefit, surely has to be fundamental to future activities related to the landscape. Economic regeneration and conservation should not only be a 'top down' process. Local communities need to be equal partners.

Above all, the Welsh landscape must be understood as a whole. Explanation, education and public participation are crucial if the rich landscape heritage of Wales is to have meaning for future generations.

Conclusion

Landscapes in the future

Jennifer Jenkins

In this small, densely populated island of Britain every landscape matters, the ordinary as much as the beautiful and the historic. During most of the twentieth century attention was concentrated on the exceptional – the National Parks, the Sites of Special Scientific Interest and the conservation areas. But as society has become more inclusive and the importance of the environment to regeneration has been demonstrated, government and public opinion have come to recognise that every single place should be valued. We can no longer tolerate the monotonous down-at-heel housing estates or the identikit red boxes sprouting on the outskirts of expanding towns. The objectives are clear: to conserve landscapes of particular beauty or historic interest; to improve decent ordinary landscapes; and to create good new landscapes where the land has been degraded. The question is how to achieve these objectives.

There has been a quantum leap in the understanding of the elements of landscape during the past decade following the Countryside Agency's pioneering work in describing the rural landscape of England and identifying areas of common character according to the landform, the present use and the historical influences. Similar studies have been initiated in Wales. If they were extended to cover the whole of Britain, they would constitute a template for reinforcing the diversity of landscapes in both town and countryside: they would counteract the trend to standardised building types and extensive suburban development unrelated to traditional patterns or indigenous plants.

Local authorities have a key role to play in protecting the best existing landscapes and fostering high-quality new development, whether in traditional or contemporary style. Unfortunately most of their planning and architecture departments no longer have the specialised staff to initiate long-term thinking or even to deal adequately with immediate problems of how to require developers to adopt higher standards of layout and design. It is essential that local authorities have the skills and the resources to provide a full planning and design service.

The public realm – the town squares and open spaces, the village greens and playgrounds – is the heart of an attractive and lively community. Here, too, insufficient local funding invariably results in low standards. Any visitor to France or Holland cannot fail to be struck by the quality of their squares and gardens, their sports grounds and even their car parks.

The Blair government has devoted considerable effort to promoting higher standards of design and encouraging an 'urban renaissance'. But no amount of rhetoric and planning guidance will be effective unless supported by financial measures that embrace both the allocation of resources and the shape of taxation. At present the reverse is true. Local authorities lack the money even to maintain their existing public parks, let alone to create a network of new open spaces; conservation is penalised by VAT that is levied on repairs but not on new building; development on greenfield sites is not only easier but usually more profitable than on brownfield sites.

We now have a chance to revive the cities, to lessen pressures on the countryside and to create new landscapes, reclaiming derelict land, extending the area of woodland and managing the changing coastline. But time is running out. Every day we lose more of our limited countryside, build more unsightly houses and offices, damage more historic skylines. Landscapes are as important to the quality of life as are the currently much discussed public services and must receive the political and financial priority they require.

Notes

Where author's name and publication date only are given, see the Bibliography for full details.

Introduction – *Jennifer Jenkins*

1 Christopher Hussey, letter to *The Times*, 25 November 1963.
2 Hilaire Belloc, *The South Country*.
3 A. G. Tansley, *Our Heritage of Wild Nature*, Readers' Union/Cambridge University Press, 1946, Cambridge.

Population pressures – *John I. Clarke*

1 The Prakesh Report (2000).
2 Shaw (2000), pp. 4–12.
3 Jackson (1998), p. 101.
4 Rowland (1979).
5 Census Research Unit (1980).
6 Denham and White (1998), pp. 23–34.
7 ibid.
8 ibid.
9 Hall (1977).
10 Denham and White (1998), pp. 23–34.
11 Shaw (2000), pp. 4–12.
12 King *et al.* (2000), pp. 13–19.
13 ibid.

14 Adams (1999).
15 Brooks (1973).

Sustainable development – *David Banister*

1 DETR (1999a).
2 Owen (2001).
3 DETR (2000a).
4 Transport's contribution to pollution in 1997 – nitrogen oxides 55% of UK total, carbon monoxide 75%, volatile organic compounds 42%, lead 61%, blacksmoke (non-burnt hydrocarbons) 60%, and particulates (PM10) 29%.
5 DETR (2000b).
6 Urban Task Force (1999), p. 254.
7 Between 1965 and 1990, some 15% of rural communities lost their last general store or food shop, and some 4,000 outlets have closed since 1990. School closures in rural areas reached a maximum level in 1983, with 127 closures, but since then the average rate of closure has been about thirty a year – only 2 rural schools closed in 1999. In the 6 months to September 2000, 299 post offices closed, but overall there have been about two hundred closures a year (all data from CAG Consultants, 2001).
8 Moseley (2000), pp. 415–34.
9 DETR (2000b, c).
10 Urban is defined here as over 10,000 people – urban areas cover 7% of the land area, with 80% of the population.
11 DETR (2000d).
12 CPRE (2001).
13 DETR (1999b).
14 Watson *et al.* (2001).
15 CAG Consultants (2001).
16 Banister and Marshall (2000).
17 DETR (2000c).
18 Banister (2000), pp. 226–35.

19 Banister (1997).

Edgelands – *Marion Shoard*

1 Dr Christine Lambert and colleagues describe the planning
  history of this area in C. Lambert, R. Griffiths, N. Oatley,
  N. Taylor and I. Smith, *On the Edge: The Development of
  Bristol's North Fringe*, Bristol Integrated Cities Study,
  Working Paper 9, 1999, University of Bristol and the
  University of the West of England.
2 Daniels (1999).
3 Lowenthal and Prince (1965), pp. 2, 185–222.
4 Ottaway (1993).
5 Salway (1993), p. 492; Toynbee (1962), pp. 37–8.
6 Hopkinson (2000).
7 Kendle and Forbes (1997), p. 40.
8 The rural white paper says, 'Farming on the urban fringe
  has its own special attributes and problems. Its landscape is
  vitally important in its own right and as a bridge to the
  wider countryside. Demand for access and amenity is high.
  Crime and vandalism can be problems.' (*Our Countryside:
  The Future: A Fair Deal for Rural England, Presented to
  Parliament by the Deputy Prime Minister and Secretary of State
  for the Environment, Transport and the Regions, and the Minister
  of Agriculture, Fisheries and Food*, Cmnd 4909, The Stationery
  Office, para. 8.61).
9 DETR (2001a), para. 3.24.
10 Countryside Commission (1999), p. 1.
11 Countryside Commission and Forestry Commission (1999).
12 Coleman (1976), pp. 3, 411–34; Coleman (1977), pp. 94–134.
13 DETR (2001b), para. 1.5.
14 For further discussion of this point see M. Shoard, 'Lie of
  the land: Marion Shoard suggests that the ban on building
  on green belt land in the south-east should be lifted',
  Environment Now, July 1988; R. Bate, 'Hitting below the

belt: Richard Bate argues that it is vital for us not to relax tight planning controls in green belt areas', *Environment Now*, September 1988; and M. Shoard, 'Lie of the land: we need a master-plan to conserve the countryside of the whole of south-east England, not just the green belt areas', *Environment Now*, November 1988.

15 Countryside Commission (1999), pp. 6–7.
16 Defoe (1962).
17 Shoard (1982).

Lowland landscapes – *Ian Hodge and Uwe Latacz-Lohmann*

1 Trist (1948), p. 134.
2 Tansley (1946), p. 63.
3 MAFF (1995).
4 The Cairns Group of exporting countries comprises Argentina, Australia, Bolivia, Brazil, Canada, Chile, Colombia, Costa Rica, Fiji, Guatemala, Indonesia, Malaysia, New Zealand, Paraguay, the Philippines, South Africa, Thailand and Uruguay.
5 Latacz-Lohmann and Hodge (2001), pp. 42–6.
6 DETR (1999).
7 Campbell *et al.* (1997).

After foot-and-mouth – *Philip Lowe*

1 Bennett *et al.* (2001).
2 Cabinet Office (2001).
3 *Farmers Weekly*, 28 April 2001.
4 HM Government (1969); Power and Harris (1973), pp. 573–96.
5 The phrase 'ultra-precautionary' was used by MAFF and DETR to characterise retrospectively MAFF's initial response to the FMD outbreak – see MAFF/DETR (2001) para. 2.

6 Countryside Agency (2001).
7 Tregear *et al.* (1998), pp. 383–94.

The farmed landscape – *Bryn Green*

1 Moore and Hooper (1975), pp. 239–50.
2 Crosby (1986).
3 Vera (2000).
4 Tallis and Switzur (1991), pp. 401–15.
5 Rackham (1986).
6 Williams (1973).
7 Scott (1942).
8 Shoard (1980), Green (1981, 1996), Body (1982), Bowers and Cheshire (1983), Harvey (1997).
9 Green (1996).
10 NCC (1984).
11 Countryside Commission (1986).
12 Potts (1971), pp. 267–71.
13 House of Commons (1997).
14 Harvey (1997); Stephenson (1997), pp. 22–6; Green (2000), pp. 253–5.
15 Bignal and McCracken (1996), pp. 413–24.
16 CEC (1988).
17 Burch *et al.* (1997).
18 Hill *et al.* (1993).
19 Colman *et al.* (1992).
20 Chamberlin (1996).
21 Ede (1998).
22 Potter and Lobley (1992).
23 CEC (1988).
24 Potter *et al.* (1991).
25 NSCGP (1992).
26 Dyson (1996).
27 Fairbrother (1970).
28 Green and Marshall (1987), pp. 143–54.

29 Hall (1990), p. 9.
30 Harms *et al.* (1993).
31 O'Riordan *et al.* (1993), pp. 123–4.
32 Colston (1997), pp. 61–7; Whitbread and Jenman (1995), pp. 84–93.
33 Green (1995).

Heritage landscapes in Wales – *Richard Keen*

1 Howells (1973).
 2 Thomas, Dylan, *Quite Early One Morning*, J. M. Dent & Sons Ltd, 1971, London, p. 1.
 3 Pennant (1883), p. 325.
 4 A key date in the history of the south Wales coalfield when Admiralty Trials resulted in the high-quality steam coals of the valleys of Glamorgan and Monmouthshire being selected as the best fuel for the British Imperial Navy.
 5 Williams (2001), p. 7.
 6 Countryside Council for Wales (November 1998).
 7 E.B. was probably Edward Bysshe, writing in *A Trip to North Wales*, published in 1702.
 8 Three registers have been published to date, dealing with Historic Parks and Gardens, Landscapes of Outstanding Historic Interest and Landscapes of Special Historic Interest.
 9 Countryside Council for Wales: Cadw (1998), p. 101.
10 Taken from Richard Llewellyn, *How Green Was My Valley*, first published in 1939.
11 National Assembly of Wales press release SDR 2/2002, 1 February 2002.
12 Extract from *National Economic Development Strategy*, European Task Force, October 1999.
13 Lovegrove *et al.* (1995).
14 HMSO (1996), pp. 46, 47, 64.
15 Report Cymru (2001).

Bibliography

Population pressures – *John I. Clarke*

Adams, J., *The Social Implications of Hypermobility*, The Environment Directorate, Organisation for Economic Co-operation and Development, 1999, Paris.

Brooks, E., *This Crowded Kingdom: An Essay on Population Pressure in Great Britain*, Knight, 1973, London.

Census Research Unit, *People in Britain. A Census Atlas*, HMSO, 1980, London.

Champion, T., Wong, C., Rooke, A., Dorling, D., Coombes, M. and Brundson, C., *The Population of Britain in the 1990s. A Social and Economic Atlas*, Clarendon Press, 1996, Oxford.

Denham, C. and White, I., 'Differences in urban and rural Britain', *Population Trends* 91, Spring 1998.

Dorling, D., 'Difficulties and dangers in estimating small area populations for health statistics', in R. Arnold, P. Elliott, J. Wakefield, and M. Quinn, (eds.), *Population Counts in Small Areas. Implications for Studies of Environment and Health*, Studies in Medical and Population Subjects, Office for National Statistics, 1999.

Hall, P., *Europe 2000*, Columbia University Press, 1977, New York.

Jackson, S., *Britain's Population: Demographic Issues in Contemporary Society*, Routledge, 1998, London.

King, D., Hayden, J., Jackson, R., Holmans, A. and Anderson, D., 'Population of households in England to 2021', *Population Trends* 99, Spring 2000.

National Statistics, *Key Population and Other Statistics, 1999.*
Series VS No. 26, PPI, No. 22.
National Statistics, *Regional Trends, 2000, UK.* No. 35, Stationery
Office, London.
National Statistics, *Social Trends, 2001, UK.* No. 31, The
Stationery Office, London.
The Prakesh Report, *The Future of Multi-Ethnic Britain*, Profile
Books/The Runnymede Trust, 2000, London.
Rowland, D.T., *Internal Migration in Australia*, Australian Bureau
of Statistics, 1979, Canberra.
Shaw, C., '1998-based national population projections for the
United Kingdom and constituent countries', *Population
Trends* 99, Spring 2000.

Sustainable development – *David Banister*

Banister, D., 'Reducing the need to travel', *Environment and
Planning* B 24 (3), 1997.
Banister, D., 'The tip of the iceberg: leisure and air travel', *Built
Environment* 26 (3), 2000.
Banister, D. and Marshall, S., *Encouraging Transport Alternatives:
Good Practice in Reducing Travel*, The Stationery Office, 2000,
London.
CAG Consultants, *State of Sustainable Development in the UK:
Central/Local Government Focus*, Report to the Sustainable
Development Commission, Final Report, February 2001,
London.
CPRE, *Running to Stand Still? An Analysis of the 10 Year Plan for
Transport*, Council for the Protection of Rural England,
February 2001, London.
DETR (1999a), *Supplementary Guidance to the Regional
Development Agencies*, DETR, April 1999, London.
DETR (1999b), *A Better Quality of Life: A Strategy for
Sustainable Development for the UK*, DETR, 1999, London.
DETR (2000a), *Transport Statistics Great Britain 2000*, The

Stationery Office, 2000, London.

DETR (2000b), *Our Towns and Cities: The Future – Delivering an Urban Renaissance*, DETR, November 2000, London.

DETR (2000c), *Our Countryside: The Future. A Fair Deal for Rural England*, DETR, November 2000, London.

DETR (2000d), *Transport 2010: The 10 Year Plan*, DETR, July 2000, London.

Moseley, M. J., 'England's village services in the late 1990s: Entrepreneurialism, community involvement and the state', *Town Planning Review* 71(4), 2000.

Owen, S., *A Strategic Analysis of the State of Sustainable Development in the UK: The Contribution Made by Regional Government, the Private Sector and NGOS, Environmental Resources Management*, January 2001, Oxford.

Urban Task Force, *Towards an Urban Renaissance*, The Stationery Office, 1999, London.

Watson, R. T., Ravindranath, N. H., Noble, I. R. and Bolin, B. (eds.), *Land Use, Land Use Change and Forestry, Special Report of the Intergovernmental Panel on Climate Change*, Cambridge University Press, 2001, Cambridge.

Edgelands – *Marion Shoard*

Coleman, A., 'Is planning really necessary?', *Geographical Journal* 142, 3, 411–434, 1976.

Coleman, A., 'Land use planning – success or failure?', *Architects' Journal* 19, January 1977.

Countryside Commission, *Linking Town and Country: Policies for the Countryside in and around Towns*, CCP 546, Countryside Commission, 1999, Cheltenham.

Countryside Commission and Forestry Commission, *What are Community Forests?* CCP 508, Countryside Commission, 1999, Cheltenham.

Daniels, T., *When City and Country Collide: Managing Growth in the Metropolitan Fringe*, Island Press, 1999, Washington DC.

Defoe, D., *A Tour through the Whole Island of Great Britain*, J. M. Dent and Sons, 1962, London.

DETR (2001a), *Planning Policy Guidance Note 7: The Countryside: Environmental Quality and Economic and Social Development,* Stationery Office, 2001, London.

DETR (2001b), *Planning Policy Guidance Note 2: Green Belts*, The Stationery Office, 2001, London.

Great North Forest, The Tees Forest, Countryside Commission, English Partnerships, *We're Transforming the Living and Working Environment of the North East*, Countryside Commission, 1999, Cheltenham.

Hopkinson, M., 'Land use survey of York's urban margins' in Hopkinson, M. (ed.), *Town and Country: Contemporary Issues at the Rural/Urban Interface*, University College of Ripon and York St John, 2000, York.

Kendle, T. and Forbes, S., *Urban Nature Conservation*, Spon Press, 1997, London.

Lowenthal, D. and Prince H., 'English landscape tastes', *Geographical Review* 55, 1965.

Ottaway, P., *English Heritage Book of Roman York*, Batsford and English Heritage, 1993, London.

Salway, P., *The Oxford Illustrated History of Roman Britain*, Oxford University Press, 1993, Oxford.

Shoard, M., 'The lure of the moors' in Gold, J. R. and Burgess, J., *Valued Environments*, Allen and Unwin, 1982, London.

Toynbee, J. M. C., *Death and Burial in the Roman World*, Clarendon Press, 1962, Oxford.

Lowland landscapes – *Ian Hodge and Uwe Latacz-Lohmann*

Campbell, L. H., Avery, M., Donald, P., Evans, A. D., Green, R. E. and Wilson, J. D., *A Review of the Indirect Effects of Pesticides on Birds*, JNCC Report 227, Joint Nature Conservation Committee, 1997, Peterborough.

DETR, *A Better Quality of Life: A Strategy for Sustainable Development*. Cmnd 4345, The Stationery Office, 1999, London.

Latacz-Lohmann, U. and Hodge, I., 'Multifunctionality and free trade – conflict or harmony?' *EuroChoices*, premier issue, 2001.

MAFF, *A Century of Agricultural Statistics. Great Britain 1866–1966*. HMSO, 1986, London.

MAFF, *European Agriculture: The Case for Radical Reform*, Minister of Agriculture, Fisheries and Food's CAP Review Group, 1995, London.

Tansley, A. G., *Our Heritage of Wild Nature*, Readers' Union/Cambridge University Press, 1946, Cambridge.

Trist, P. J. O., *Land Reclamation*, Faber and Faber, 1948, London.

After foot-and-mouth – *Philip Lowe*

Bennett, K., Phillipson, J., Lowe, P. and Ward, N., *The Impact of the Foot and Mouth Crisis on Rural Firms: A Survey of Microbusinesses in the North East of England*, Centre for Rural Economy Research Report, 2001, University of Newcastle upon Tyne.

Cabinet Office, *Comparing the 1967–8 and the 2001 Foot and Mouth Epidemics*, 2001 (http://www.maff.gov.uk/animalh/ diseases/fmd/disease/comparison/introduction.asp).

Countryside Agency, *State of the Countryside 2001*, Countryside Agency, 2001, Cheltenham.

Harvey, D., *What Lessons from Foot and Mouth? A Preliminary Economic Assessment of the 2001 Epidemic*, Centre for Rural Economy Working Paper 63, 2001, University of Newcastle upon Tyne.

HM Government, *The Report of the Committee of Inquiry on Foot and Mouth Disease, Parts 1 and 2* (the 'Northumberland Report'), HMSO, 1969, London.

MAFF/DETR, *Guidance for Local Authorities in England on*

Public Access to the Countryside on the Rights of Way Network, 28 May 2001.

Power, A. and Harris, S., 'A cost-benefit evaluation of alternative control policies for foot and mouth disease in Great Britain', *Journal of Agricultural Economics* 24, 1973.

Tregear, A., Kuznesof, S. and Moxey, A., 'Policy initiatives for regional foods: some insights from consumer research', *Food Policy* 23 (5), 1998.

The farmed landscape – *Bryn Green*

Bignal, E. M. and McCracken, D. I., 'Low-intensity farming systems in the conservation of the countryside', *Journal of Applied Ecology* 33, 1996.

Body, R., *Agriculture: The Triumph and the Shame*, Temple Smith, 1982, London.

Bowers, J. K. and Cheshire, P., *Agriculture, the Countryside and Land Use*, Methuen, 1983, London.

Burch, F. M., Green, B. H., Mitchley, J. and Potter, C. A., 'Possible options for the better integration of environmental concerns into support for arable crops', Research report to the European Commission, 1997, Wye College, University of London.

Campbell, L. H. and Cooke, A. S. (eds.), *The Indirect Effects of Pesticides on Farmland Birds*, Joint Nature Conservation Committee, 1997, London.

CEC, *The Future of Rural Society*, Com (88) 501 final. Commission of the European Communities, 1988, Brussels.

Chamberlin, B., *Farming and Subsidies*, Euroa Farms Ltd, 1996, Pukekohe, New Zealand.

Colman, D., Crabtree, J., Froud, J. and O'Carroll, L., *Comparative Effectiveness of Conservation Mechanisms*, Dept. Ag. Econ., University of Manchester, 1992, Manchester.

Colston, A., 'Conserving wildlife in a black hole', *Ecos* 18, 1997.

Countryside Commission, *Monitoring Landscape Change*, Countryside Commission, 1986, Cheltenham.

Crosby, A. W., *Ecological Imperialism*, Cambridge University Press, 1986, Cambridge.

DETR, *Digest of Environmental Statistics* 20, HMSO, 1998, London.

Dyson, T., *Population and Food. Global Trends and Future Prospects*, Routledge, 1996, London.

Ede, J., 'Every little helps... What do arable farm assurance schemes offer wildlife?' *Ecos* 19, 1998.

Fairbrother, N., *New Lives, New Landscapes*, Architectural Press, 1970, London.

Green, B., *Countryside Conservation*, E. & F. N. Spon, 1981, revised 1996, London.

Green, B., 'Plenty and wilderness? Creating a new countryside', *Ecos* 16, 1995.

Green, B. H., 'Farming the environment – lessons from New Zealand', *Town and Country Planning* 69, 2000.

Green, B. H. and Marshall, I. C., 'An assessment of the role of golf courses in Kent, England, in protecting wildlife and landscapes', *Landscape and Urban Planning* 14, 1987.

Hall, P., 'Bring back the Parkway', *Planner* 76, 1990.

Harms, B. H., Knappen, J. P. and Rademakers, J. G., 'Landscape planning for nature restoration: comparing regional scenarios' in Vos, C. C. and Opdam, P. (eds.), *Landscape Ecology of a Stressed Environment*, Chapman and Hall, 1993, London.

Harvey, G., *The Killing of the Countryside*, Cape, 1997, London.

Hill, P., Green, B. H. and Edwards, A., *The Cost of Care*, RICS, 1993, London.

House of Commons, *Environmentally Sensitive Areas and Other Schemes under the Agri-environment Regulation*, HMSO, 1997, London.

Moore, N. W. and Hooper, M. D., 'On the number of bird species in British woods', *Biological Conservation* 8, 1975.

NCC, *Nature Conservation in Great Britain*, Nature Conservancy Council, 1984, Peterborough.

NSCGP, *Ground for Choices: Four Perspectives for the Rural Areas in the European Community*, Netherlands Scientific Council for Government Policy, 1992, The Hague.

O'Riordan, T., Wood, C. and Sheldrake, A., 'Landscapes for tomorrow', *Journal of Environmental Planning* 36, 1993.

Potter, C., Burnham, P., Edwards, A., Gasson, R. and Green, B. H., *The Diversion of Land: Conservation in a Period of Farming Contraction*, Routledge, 1991, London.

Potter, C. and Lobley, M., 'The conservation status and potential of elderly farmers; results from a survey in England and Wales', *Journal of Rural Studies* 8, 1992.

Potts, G. R., 'Agriculture and the survival of partridges', *Outlook Agriculture* 6, 1971.

Rackham, O., *The History of the Countryside*, Temple Smith, 1986, London.

Scott, Lord Justice, 'Report of the committee on land utilisation in rural areas', Cmnd 6378, Ministry of Works and Planning, HMSO, 1942, London.

Shoard, M., *The Theft of the Countryside*, Temple Smith, 1980, London.

Stephenson, G., 'Is there life after subsidies? The New Zealand experience', *Ecos* 18, 1997.

Tallis, J. and Switzur, V. R., 'The spread of moorland – local, regional and national', *Journal of Ecology* 79, 1991.

Vera, F. W. M., *Grazing Ecology and Forest History*, CABI Publishing, 2000, Wallingford.

Whitbread, A. and Jenman, W., 'A natural method of conserving biodiversity in Britain', *British Wildlife* 7, 1995.

Williams, R., *The Country and the City*, Chatto & Windus, 1973, London.

Heritage landscapes in Wales – *Richard Keen*

Countryside Council for Wales in association with the Wales Landscape Partnership Group, *LANDMAP. The Landscape*

Assessment and Decision-making Process, draft handbook for consultation, November 1998.

Countryside Council for Wales, Cadw: Welsh Historic Monuments, ICOMOS UK, *Landscape of Outstanding Historic Interest in Wales*, Part 2.1, 1998, Cardiff.

HMSO, *A Working Countryside for Wales*, 1996, London.

Howells, B. E. (ed.), *Elizabethan Pembrokeshire: The Evidence of George Owen*, Pembrokeshire Record Society, 1973, Haverfordwest.

Lovegrove, R., Shrubb, M. and Williams, I., *Silent Fields Gwlad Tawel: The Current Status of Farmland Birds in Wales*, RSPB, 1995, Bedfordshire.

Pennant, T., *A Tour in Wales* (abridged by David Kirk), Gwasg Carreg Gwalch, Llanwrst, 1998, p. 103.

Report Cymru, Heritage Lottery Fund and Welsh Development Agency, *Valuing Our Environment: Economic Impact of the Environment of Wales*, July 2001.

Williams, M., 'Landscape or sustainability? Is there a choice?', *Rural Wales*, Spring 2001.

Acknowledgements

John I. Clarke acknowledges his debt to the Office of National Statistics for extensive use of data derived from its quarterly publication *Population Trends* and its annual publications *Regional Trends, Social Trends, Key Population and Vital Statistics* and *International Migration*, and particularly to its website and those of the Scottish Executive and the General Register Office for Scotland.

Richard Keen wishes to thank Richard Kelly of the Countryside Council for Wales for the information on Tir Cymen and Tir Gofal and other valuable comments and advice.

Philip Lowe wishes to thank Christopher Ray for his permission to reproduce his photographs taken in rural Northumberland.

Sir Crispin Tickell wishes to thank the Natural History Museum, Dr Antony Sutcliffe and the Centre for Ecology and Hydrology for their assistance in providing illustrations.

Index

Photographs in the plate sections are denoted by *italic* plate numbers. Figures in the text are indicated by *italic* page numbers.